Springer Undergraduate Mathematics Series

The Springer Undergraduate Mathematics Series (SUMS) is a series designed for undergraduates in mathematics and the sciences worldwide. From core foundational material to final year topics, SUMS books take a fresh and modern approach. Textual explanations are supported by a wealth of examples, problems and fully-worked solutions, with particular attention paid to universal areas of difficulty. These practical and concise texts are designed for a one- or two-semester course but the self-study approach makes them ideal for independent use.

More information about this series at http://www.springer.com/series/3423

Shoshichi Kobayashi

Differential Geometry of Curves and Surfaces

Springer

Shoshichi Kobayashi (1932–2012)
University of California
Berkeley, USA

Translated by
Eriko Shinozaki Nagumo
Kanagawa, Japan

Makiko Sumi Tanaka
Tokyo University of Science
Chiba, Japan

ISSN 1615-2085 ISSN 2197-4144 (electronic)
Springer Undergraduate Mathematics Series
ISBN 978-981-15-1738-9 ISBN 978-981-15-1739-6 (eBook)
https://doi.org/10.1007/978-981-15-1739-6

Mathematics Subject Classification (2010): 53A04, 53A05, 53A10

This Springer imprint is published by the registered company Springer Nature Singapore Pte Ltd.
The registered company address is: 152 Beach Road, #21-01/04 Gateway East, Singapore 189721, Singapore

Preface

This book was originally published in 1977 under the same title as one of the "Kiso Sugaku Sensho" (Selection of Fundamental Mathematics) series by Shokabo Co., Ltd. After twenty-two printings, it has been completely revised and a new chapter "Minimal Surfaces" has been added.

In Chapter 1 through Chapter 4, all detected errors have been corrected and some sentences have been changed, but the book should still be accessible to readers with knowledge of elementary calculus and matrices of size 2 and 3 only, which was the case with the original edition. In order to understand the new chapter "Minimal Surfaces," the readers should be familiar with elementary complex function theory. To preserve the introductory nature of the book, we do not refer to difficult results in analysis, but focus instead on studying further details of examples of minimal surfaces introduced in Chapter 2.

In this revised edition we made a special effort to improve the illustrative figures. We have added computer generated figures of some minimal surfaces, which should help the readers better visualize the surfaces. These figures were produced by Ms. Junko Matsuoka, a graduate student of Tsuda College.

Furthermore, I am greatly indebted to Mr. Shuji Hosoki and Ms. Seiko Nomura of Shokabo for the publication of this revised edition. My former teacher Professor Kentaro Yano, to whom I owed the first edition of this book, sadly passed away two years ago. May his soul rest in peace.

Summer 1995

Shoshichi Kobayashi

Preface to the First Edition

This book is written for readers who have studied college-level calculus and matrices of order 2 or 3. The purpose of this book is to have such beautiful and profound results in geometry as the Gauss-Bonnet theorem understood by readers who have not yet learned much advanced mathematics. Therefore, this book does not discuss as many results as other books on the theory of curves and surfaces, but provides detailed explanations of several so-called global results. Ideally, it might have been better to have written a little more about minimal surfaces, which are actively studied even today, but in order to keep the book at an introductory level, we discuss only a few examples. Those readers who become interested in differential geometry after reading this book are advised to study manifolds, which are surfaces of higher dimensions, and then differential geometry on manifolds, rather than to study further details of curve theory and surface theory. It may be good to examine the theory of surfaces once again after this book. Nowadays, many universities teach manifolds but not even elementary differential geometry. I would like students in such universities to study by themselves the theory of curves and surfaces, at least at the level treated in this book. Thus, we have provided solutions to all exercise problems for the benefit of such readers.

Lastly, I would like to thank my former teacher, Professor Kentaro Yano, who suggested me to write this book and has advised me over many years.

I have not written a mathematics book in Japanese before this. My poorly prepared manuscript would not have reached the stage of its publication were it not for the great help from Mr. Kyohei Endo and Mr. Shuji Hosoki of Shokabo, as well as Professor Yano. It is not uncommon for a teacher to place a burden on his student in writing a book, but this book is unique in that the student troubled much his teacher.

Berkeley, Summer 1977

Shoshichi Kobayashi

Translators' Note

I first met Professor Shoshichi Kobayashi 20 years ago, during my junior year of college, when he was a visiting professor at International Christian University (Tokyo, Japan). After taking his class, I became interested in continuing my study of differential geometry and wished to write a thesis on this topic. Since there was no differential geometry professor at ICU, Professor Kobayashi kindly advised me to continue my differential geometry studies under Ms. Makiko Tanaka (currently a professor at Tokyo University of Science), who was a fixed-term lecturer at ICU at the time. She had earned her doctorate in Mathematics at Sophia University in differential geometry, under the supervision of Professor Tadashi Nagano, who was an old friend of Professor Kobayashi.

Professor Kobayashi also offered me the opportunity to continue my differential geometry studies under his guidance. He helped me to choose a thesis topic on conjugate connections. We proved a theorem together and published a joint paper in "Tokyo Journal of Mathematics" in 1997. However, since I decided to work full time as a mathematics teacher in a Japanese high school, rather than pursue graduate studies (finding a job was very competitive in Japan in 1995), it became difficult for me to continue my research. Then one day, Professor Kobayashi asked me to help him translate and type his Japanese book "Differential Geometry of Curves and Surfaces" in English using LaTeX. Since he wanted me to solve all of the problems and review differential geometry topics when translating, it took us 20 years to translate up to the end of Chapter 3.

I would like to thank Professor Kobayashi with all my heart for giving me the opportunity to continue working in the field of mathematics with him and to express my appreciation for his confidence in my ability to translate and understand his work. I would also like to thank Professor Makiko Sumi Tanaka, who always kindly kept in touch with me and helped me to finish this translation.

Eriko Shinozaki Nagumo

After Professor Shoshichi Kobayashi passed away in August of 2012, Ms. Eriko Nagumo, my former student, talked to me about her long-term project with Professor Kobayashi to translate his Japanese book into English. I believe that "Differential Geometry of Curves and Surfaces" is an excellent introductory differential geometry book, and as Ms. Nagumo was uncertain of how to proceed with the project, I offered my assistance in having the English translation published. It would be my great pleasure if Professor Kobayashi would find our work acceptable.

I have also contributed to making additions to the proof of the special case of Fenchel's theorem (Theorem 1.3.2), in which we noticed an incompleteness in the argument in Japanese edition. For readers' convenience, I have also provided some relatively new references in this field:

1. M. P. Do Carmo, Differential Geometry of Curves and Surfaces: Revised and Updated Second Edition, Dover, 2016.
2. A. Gray, E. Abbena and S. Salamon, Modern Differential Geometry of Curves and Surfaces with MATHEMATICA (3rd edition), Chapman and Hall, 2006.
3. S. Montiel and A. Ros, Curves and Surfaces, Second Edition, AMS, 2009.
4. M. Umehara and K. Yamada, Differential Geometry of Curves and Surfaces, Translated by Wayne Rossman, World Scientific, 2017.

Makiko Sumi Tanaka

Acknowledgments: The translators would like to thank Mr. Masayuki Nakamura of Springer and Mr. Toshiteru Kojima of Shokabo for their assistance in publishing this translated version. They also thank Prof. Hisashi Kobayashi, Prof. Brian L. Mark, Dr. Linda M. Zeger and Dr. Kurando Baba for their editorial comments and suggestions, and Mrs. Yukiko Kobayashi for her long time support and encouragements.

Contents

Chapter 1
Plane Curves and Space Curves

1.1 Concept of Curves

If we draw the graph of a function $y = f(x)$ on the plane, we obtain a curve. For example, the graph of $y = x^2$ is a parabola.

Similarly, the graph of a function $x = g(y)$, for example $x = y^2$, is also a curve. Both $y = f(x)$ and $x = g(y)$ have one of the variables as an independent variable, and the other as a dependent variable. So x and y are not equally treated. If we rewrite these in the form $y - f(x) = 0$ or $x - g(y) = 0$, we can unify them in the form

$$F(x, y) = 0,$$

and x and y are now treated equally. The equation $F(x, y) = 0$ expresses a more general curve. The equation of a circle, $x^2 + y^2 - a^2 = 0$ has this form. We can of course write this as $y = \pm\sqrt{a^2 - x^2}$, but this will require two equations, and its form is not beautiful. Also, it is difficult to rewrite equations such as $x^3 + x^2y + y^3 = 0$ in the form $y = f(x)$. However, an equation $F(x, y) = 0$ does not always define a curve. For example, $x^2 + y^2 + 1 = 0$ gives an empty set. An equation of the form $F(x, y) = 0$ is useful when variables x and y are not real variables, but complex variables, and represents a polynomial of x and y. However curves of this type are studied in algebraic geometry rather than in differential geometry.

In differential geometry, the most useful way to express a curve is by "parametric representation". Let

$$x = x(t), \qquad y = y(t) \tag{1.1.1}$$

be functions of a real variable t. When t moves within an interval $a \leq t \leq b$, $(x(t), y(t))$ draws a curve on the plane. The curve $y = f(x)$ mentioned before, can be written as $x = t$, $y = f(t)$, so it can be parametrically represented. Actually, it may be difficult to express the curve given by $F(x, y) = 0$ in the form (1.1.1). However circles like $x^2 + y^2 - a^2 = 0$ can be written as

$$x = a\cos t, \qquad y = a\sin t. \tag{1.1.2}$$

The original version of the chapter was revised: Belated corrections have been incorporated. The correction to the chapter is available at https://doi.org/10.1007/978-981-15-1739-6_6

S. Kobayashi, *Differential Geometry of Curves and Surfaces*,
Springer Undergraduate Mathematics Series,
https://doi.org/10.1007/978-981-15-1739-6_1

Generally, when we are given a curve by (1.1.1), it will be easier to understand if we consider the point $(x(t), y(t))$ as a particle moving with time t. In differential geometry, we assume that functions $x(t)$ and $y(t)$ are not only continuous, but also differentiable in t, at least two or three times. Sometimes we assume that $x(t), y(t)$ are not differentiable at some points (intuitively speaking, the curve has edges) but we can understand such an assumption from the context even if it is not stated explicitly.

We have been talking about plane curves so far, while space curves are given by three functions,

$$x = x(t), \qquad y = y(t), \qquad z = z(t). \tag{1.1.3}$$

More generally, a curve in an n-dimensional Euclidean space is written as

$$x^1 = x^1(t), \quad x^2 = x^2(t), \quad \dots, \quad x^n = x^n(t). \tag{1.1.4}$$

(Here, we denote coordinates of the n-dimensional space with $(x^1, x^2, \cdots, x^n)$, not with $(x_1, x_2, \cdots, x_n)$. This is in conformity with tensor analysis, so we need to be careful not to confuse x^2 with x squared.) In this chapter, we only consider curves in two or three dimensional space but it will be easier if we write (1.1.4) using the vector notation

$$\boldsymbol{p} = \boldsymbol{p}(t). \tag{1.1.5}$$

1.2 Plane Curves

We set plane curves using coordinate system (x, y),

$$x = x(t), \qquad y = y(t). \tag{1.2.1}$$

As in (1.1.5), we will also write it as

$$\boldsymbol{p} = \boldsymbol{p}(t). \tag{1.2.2}$$

We differentiate (1.2.1) and (1.2.2) and write it as

$$\dot{\boldsymbol{p}}(t) = (\dot{x}(t), \dot{y}(t)). \tag{1.2.3}$$

The custom of writing a derivative as $\dot{x}$ instead of x' or dx/dt, originates from physics. If we consider $\boldsymbol{p}(t)$ as a point at time t, $\dot{\boldsymbol{p}}(t)$ is the velocity vector of a moving point. The length $|\dot{\boldsymbol{p}}(t)|$ of a vector $\dot{\boldsymbol{p}}(t)$ is given by

$$|\dot{\boldsymbol{p}}(t)| = \sqrt{(\dot{x}(t))^2 + (\dot{y}(t))^2}, \tag{1.2.4}$$

and it represents the speed of the motion. Similarly, the vector

$$\ddot{\boldsymbol{p}}(t) = (\ddot{x}(t), \ddot{y}(t)) \tag{1.2.5}$$

which is obtained by differentiating $\dot{\boldsymbol{p}}(t)$ represents the acceleration vector.

The length of the curve $\boldsymbol{p}(t)$ $(0 \leq t \leq b)$ is given by

$$\int_0^b |\dot{\boldsymbol{p}}(t)| dt.$$

This is the distance traveled by $\boldsymbol{p}(t)$ between time $t = 0$ and $t = b$. By fixing the initial time 0 and changing the variable from b to t,

$$s = \int_0^t |\dot{\boldsymbol{p}}(t)| dt. \tag{1.2.6}$$

Then s is the distance the point moved between time 0 and t. Thus it obviously is a function of t. To be precise, we should write it as

$$s = \int_0^t |\dot{\boldsymbol{p}}(u)| du$$

to distinguish t from u but it is customary to write as (1.2.6). Conversely, consider the problem of writing t as a function of s. Physically speaking, if the moving point does not stop between 0 and t, at least theoretically it is possible to determine t from the distance s. Mathematically, the fundamental theorem of calculus applied to (1.2.6) gives

$$\dot{s}(t) = |\dot{\boldsymbol{p}}(t)|. \tag{1.2.7}$$

If $\dot{\boldsymbol{p}}(t) \neq \boldsymbol{0}$, (so that $|\dot{\boldsymbol{p}}(t)| > 0$) for all t, then s will be a monotone increasing function of t, and it is possible to write t as a function of s. Thus by substituting $t = t(s)$ into (1.2.1) or (1.2.2), we can express the curve using the length parameter s.

For simplicity, we assume that (1.2.1) and (1.2.2) are already written using the parameter s,

$$\boldsymbol{p} = \boldsymbol{p}(s) = (x(s), y(s)). \tag{1.2.8}$$

It is customary to write

$$\boldsymbol{p}' = \boldsymbol{p}'(s) = (x'(s), y'(s)) \tag{1.2.9}$$

when we differentiate it by s, which is the usual notation of differentiation. Since

$$\frac{dx}{ds} = \frac{dx}{dt}\frac{dt}{ds}, \qquad \frac{dy}{ds} = \frac{dy}{dt}\frac{dt}{ds},$$

using (1.2.7), we obviously get

$$|\boldsymbol{p}'(s)| = \sqrt{\left(\frac{dx}{dt}\right)^2 + \left(\frac{dy}{dt}\right)^2}\frac{dt}{ds} = \frac{ds}{dt}\frac{dt}{ds} = 1. \tag{1.2.10}$$

In physics, this indicates that movement (1.2.8) has constant speed 1. Let $\boldsymbol{e}_1 = \boldsymbol{e}_1(s)$ be the vector $\boldsymbol{p}' = \boldsymbol{p}'(s)$ with length 1 given by (1.2.9). Thus,

$$e_1 = p'. \tag{1.2.11}$$

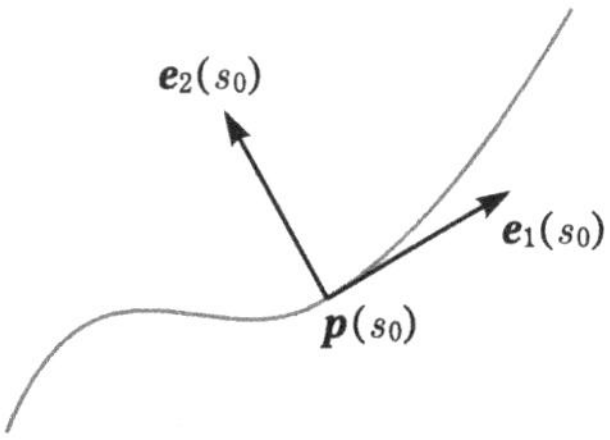

Fig. 1.2.1

Geometrically, $e_1(s_0)$ indicates the tangent vector of a curve $p(s)$ at $p(s_0)$. The equation of the tangent line at that point is,

$$p(s) = p(s_0) + e_1(s_0)(s - s_0), \tag{1.2.12}$$

or

$$\begin{cases} x = x(s_0) + x'(s_0)(s - s_0), \\ y = y(s_0) + y'(s_0)(s - s_0). \end{cases} \tag{1.2.13}$$

We get (1.2.12) by expanding $p(s)$ in a power series of s at s_0 and then discarding the terms of second order or higher. In other words, (1.2.12) is the first order approximation of (1.2.8) at s_0.

Let $e_2(s_0)$ be a vector perpendicular to $e_1(s_0)$, with length 1. There are two choices for $e_2(s_0)$, but choose the one pointing to the left of $e_1(s_0)$. The following example is important in order to understand the definition of curvature.

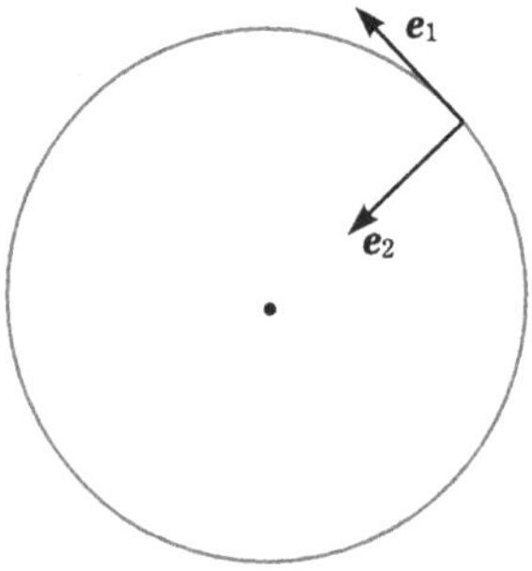

Fig. 1.2.2

Example 1.2.1

$$x(t) = r\cos t, \qquad y(t) = r\sin t \tag{1.2.14}$$

is the equation of a circle with radius r. In order to rewrite it with parameter s as in (1.2.8), we calculate (1.2.6) to obtain

$$s = \int_0^t \sqrt{\dot{x}(t)^2 + \dot{y}(t)^2}dt = \int_0^t r dt = rt. \tag{1.2.15}$$

Thus, the equation of a circle in terms of the parameter s is given by

$$x(s) = r\cos\frac{s}{r}, \qquad y(s) = r\sin\frac{s}{r}. \tag{1.2.16}$$

Differentiating this by s we get,

$$x'(s) = -\sin\frac{s}{r}, \qquad y'(s) = \cos\frac{s}{r}, \tag{1.2.17}$$

$$\boldsymbol{e}_1(s) = \left(-\sin\frac{s}{r},\ \cos\frac{s}{r}\right). \tag{1.2.18}$$

Thus,

$$\boldsymbol{e}_2(s) = \left(-\cos\frac{s}{r},\ -\sin\frac{s}{r}\right). \tag{1.2.19}$$

By differentiating $\boldsymbol{e}_1$ and $\boldsymbol{e}_2$ by s we get,

$$\begin{aligned} \boldsymbol{e}_1'(s) &= \left(-\frac{1}{r}\cos\frac{s}{r},\ -\frac{1}{r}\sin\frac{s}{r}\right) = \frac{1}{r}\,\boldsymbol{e}_2(s), \\ \boldsymbol{e}_2'(s) &= \left(\frac{1}{r}\sin\frac{s}{r},\ -\frac{1}{r}\cos\frac{s}{r}\right) = -\frac{1}{r}\,\boldsymbol{e}_1(s). \end{aligned} \tag{1.2.20}$$

♦

In general, the relationship between the tangent vector $\boldsymbol{e}_1(s)$ of a curve $\boldsymbol{p}(s)$ and the vector $\boldsymbol{e}_2(s)$ perpendicular to $\boldsymbol{e}_1(s)$ can be expressed as follows,

$$\boldsymbol{p}' = \boldsymbol{e}_1, \quad \boldsymbol{e}_1\cdot\boldsymbol{e}_1 = 1, \quad \boldsymbol{e}_2\cdot\boldsymbol{e}_2 = 1, \quad \boldsymbol{e}_1\cdot\boldsymbol{e}_2 = 0. \tag{1.2.21}$$

The symbol $\cdot$ in $\boldsymbol{e}_1\cdot\boldsymbol{e}_1$, $\boldsymbol{e}_2\cdot\boldsymbol{e}_2$, and $\boldsymbol{e}_1\cdot\boldsymbol{e}_2$ represents the inner product.

By differentiating $\boldsymbol{e}_1\cdot\boldsymbol{e}_1 = 1$, we get $\boldsymbol{e}_1'\cdot\boldsymbol{e}_1 + \boldsymbol{e}_1\cdot\boldsymbol{e}_1' = 0$. Hence, $2\boldsymbol{e}_1'\cdot\boldsymbol{e}_1 = 0$. Therefore $\boldsymbol{e}_1'$ is perpendicular to $\boldsymbol{e}_1$. A vector perpendicular to $\boldsymbol{e}_1(s)$ is a multiple of $\boldsymbol{e}_2(s)$, so we get

$$\boldsymbol{e}_1'(s) = \kappa(s)\,\boldsymbol{e}_2(s) \tag{1.2.22}$$

(by setting the multiple to $\kappa(s)$). Similarly, by differentiating $\boldsymbol{e}_2\cdot\boldsymbol{e}_2 = 1$, we see that $\boldsymbol{e}_2'$ is perpendicular to $\boldsymbol{e}_2$, and $\boldsymbol{e}_2'$ is a multiple of $\boldsymbol{e}_1$. On the other hand, differentiating $\boldsymbol{e}_1\cdot\boldsymbol{e}_2 = 0$ we get $\boldsymbol{e}_1'\cdot\boldsymbol{e}_2 + \boldsymbol{e}_1\cdot\boldsymbol{e}_2' = 0$. Substituting $\boldsymbol{e}_1' = \kappa\boldsymbol{e}_2$ and using $\boldsymbol{e}_2\cdot\boldsymbol{e}_2 = 1$, we get $\kappa + \boldsymbol{e}_1\cdot\boldsymbol{e}_2' = 0$, so $\boldsymbol{e}_2' = -\kappa\boldsymbol{e}_1$. Thus we have

$$\begin{cases} e_1' &= \quad \kappa e_2, \\ e_2' &= -\kappa e_1, \end{cases} \tag{1.2.23}$$

which can also be written as

$$\begin{cases} e_1' &= 0\, e_1 + \kappa e_2, \\ e_2' &= -\kappa e_1 + 0\, e_2. \end{cases}$$

Note that the following matrix

$$\begin{bmatrix} 0 & \kappa \\ -\kappa & 0 \end{bmatrix}, \tag{1.2.24}$$

made from coefficients of the above equation, is an alternating matrix. Note also that for the circle with radius r in Example 1.2.1, $\kappa = \frac{1}{r}$. We call this $\kappa(s)$ the **curvature** of the curve $\boldsymbol{p}(s)$. If the curvature is identically 0, in other words $\kappa(s) \equiv 0$, from (1.2.22) we get $\boldsymbol{e}_1' = \boldsymbol{0}$, and we know that $\boldsymbol{e}_1(s) = (x'(s), y'(s))$ is a constant vector independent of s. Let $x'(s) = \alpha$, $y'(s) = \beta$. Then,

$$x(s) = \alpha s + a, \qquad y(s) = \beta s + b$$

for constants a, b, which is the equation of a line. Hence, we have

Theorem 1.2.1 *The curvature of $\boldsymbol{p}(s)$ is identically* 0, *if and only if $\boldsymbol{p}(s)$ is a line.*

From Example 1.2.1 and Theorem 1.2.1, we can see that "curvature" shows how sharply a curve is curving. To be more precise, we continue expanding (1.2.12). Then we have

$$\boldsymbol{p}(s) = \boldsymbol{p}(s_0) + \boldsymbol{p}'(s_0)(s - s_0) + \frac{1}{2}\boldsymbol{p}''(s_0)(s - s_0)^2 + \cdots . \tag{1.2.25}$$

Using the relation $\boldsymbol{p}' = \boldsymbol{e}_1$ and $\boldsymbol{p}'' = \boldsymbol{e}_1' = \kappa \boldsymbol{e}_2$, we rewrite it as follows:

$$\boldsymbol{p}(s) = \boldsymbol{p}(s_0) + \boldsymbol{e}_1(s_0)(s - s_0) + \frac{1}{2}\kappa(s_0)\boldsymbol{e}_2(s_0)(s - s_0)^2 + \cdots . \tag{1.2.26}$$

By discarding the terms indicated by "$\cdots$" we get the second order approximation of $\boldsymbol{p}(s)$ at s_0.

Another way of explaining curvature is by "Gauss map." To each point $\boldsymbol{p}(s)$ on a curve $\boldsymbol{p}$ there corresponds a vector at the origin parallel to $\boldsymbol{e}_2(s)$.

Since this parallel vector at the origin is the same as $\boldsymbol{e}_2(s)$ at the point $\boldsymbol{p}(s)$, we write it also as $\boldsymbol{e}_2(s)$. We write the correspondence $\boldsymbol{p}(s) \to \boldsymbol{e}_2(s)$ from the curve $\boldsymbol{p}$ to the unit circle as g taking the initial g of "Gauss." Now, as a point on the curve $\boldsymbol{p}$ moves from $\boldsymbol{p}(s)$ to $\boldsymbol{p}(s + \Delta s)$ by a small distance Δs, $\boldsymbol{e}_2$ moves from $\boldsymbol{e}_2(s)$ to $\boldsymbol{e}_2(s + \Delta s)$. Since

$$\boldsymbol{e}_2(s + \Delta s) = \boldsymbol{e}_2(s) + \boldsymbol{e}_2'(s)\Delta s + \cdots = \boldsymbol{e}_2(s) - \kappa(s)\boldsymbol{e}_1(s)\Delta s + \cdots , \tag{1.2.27}$$

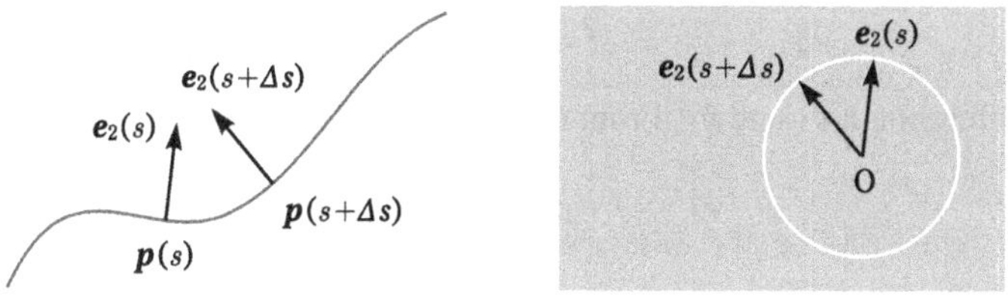

Fig. 1.2.3

$g(\boldsymbol{p})$ moves by distance $\kappa(s)\Delta s$ in the direction of $-\boldsymbol{e}_1(s)$, i.e., anti-clockwise on the unit circle. Thus, as $\boldsymbol{p}$ moves with speed 1, $g(\boldsymbol{p})$ moves on the unit circle with speed κ. When κ is negative, $g(\boldsymbol{p})$ moves in the negative direction, i.e., clockwise. The reader who finds understanding the direction in which $g(\boldsymbol{p})$ moves difficult, may look at Figure 1.2.3.

We have already checked in Example 1.2.1 that the circle with radius r has a curvature $\frac{1}{r}$. Now let us prove that if the curvature of a curve $\boldsymbol{p}$ is constant $\kappa > 0$, then $\boldsymbol{p}$ describes a circle with radius $\frac{1}{\kappa}$. By looking at Example 1.2.1, we can guess that the center of this circle is $\boldsymbol{p} + \frac{1}{\kappa}\boldsymbol{e}_2$. First, in order to check that $\boldsymbol{p} + \frac{1}{\kappa}\boldsymbol{e}_2$ is a constant point not depending on s, we differentiate it by s. Thus

$$\boldsymbol{p}' + \frac{1}{\kappa}\boldsymbol{e}_2' = \boldsymbol{e}_1 + \frac{1}{\kappa}(-\kappa\boldsymbol{e}_1) = \boldsymbol{0}. \tag{1.2.28}$$

Now, we know that $\boldsymbol{p} + \frac{1}{\kappa}\boldsymbol{e}_2$ is a constant point. Since the vector from this constant point $\boldsymbol{p} + \frac{1}{\kappa}\boldsymbol{e}_2$ to $\boldsymbol{p}(s)$ is given by $\boldsymbol{p}(s) - (\boldsymbol{p}(s) + \frac{1}{\kappa}\boldsymbol{e}_2) = -\frac{1}{\kappa}\boldsymbol{e}_2$ and since it has length $|-\frac{1}{\kappa}| = \frac{1}{\kappa}$, we know that $\boldsymbol{p}(s)$ is located on the circle with center $\boldsymbol{p} + \frac{1}{\kappa}\boldsymbol{e}_2$ and radius $\frac{1}{\kappa}$. The reader should check that for a negative constant κ, $\boldsymbol{p}$ moves on a circle with radius $\frac{1}{|\kappa|}$ in the negative direction, i.e., clockwise.

What we have proved so far is a special case of the general fact that a curve on the plane is determined by its curvature. Before proving this fact, we note that the curvature of two congruent curves are equal to each other, that is, the curvature does not change when a curve $\boldsymbol{p}$ is rotated and displaced in parallel on the plane. This can be seen using the Gauss map, but in order to check it with calculation, rotate $\boldsymbol{p}$ by an orthogonal matrix A of size 2 and then displace it in parallel by a vector $\boldsymbol{a}$ so that the resulting curve $\bar{\boldsymbol{p}}$ is given by

$$\bar{\boldsymbol{p}} = A\boldsymbol{p} + \boldsymbol{a}, \tag{1.2.29}$$

(where we consider $\boldsymbol{p}$ in $A\boldsymbol{p}$ as column vector, and so forth). From this we get

$$\bar{\boldsymbol{e}}_1 = \bar{\boldsymbol{p}}' = A\boldsymbol{p}' = A\boldsymbol{e}_1. \tag{1.2.30}$$

Since this shows that $\bar{\boldsymbol{e}}_1$ is obtained from $\boldsymbol{e}_1$ by the rotation A, $\bar{\boldsymbol{e}}_2$ is also obtained from $\boldsymbol{e}_2$ by the same rotation A, that is,

$$\bar{\boldsymbol{e}}_2 = A\boldsymbol{e}_2. \tag{1.2.31}$$

Next we differentiate $\boldsymbol{e}_1$ and $\bar{\boldsymbol{e}}_1$. From (1.2.30), (1.2.31), and

$$\boldsymbol{e}_1{}' = \kappa\,\boldsymbol{e}_2, \qquad \bar{\boldsymbol{e}}_1' = \bar{\kappa}\,\bar{\boldsymbol{e}}_2. \tag{1.2.32}$$

It follows that $\kappa = \bar{\kappa}$. Thus, because the curvature does not change by congruent transformation, it is truly a geometric quantity.

Let us now prove that when the two curvatures $\kappa(s)$ and $\bar{\kappa}(s)$ of curves $\boldsymbol{p}(s)$ and $\bar{\boldsymbol{p}}(s)$ are equal, we get $\bar{\boldsymbol{p}}$ from $\boldsymbol{p}$ by rotation and parallel displacement. We apply appropriate rotation and parallel displacement to $\boldsymbol{p}$ so that

$$\boldsymbol{p}(s_0) = \bar{\boldsymbol{p}}(s_0), \qquad \boldsymbol{e}_1(s_0) = \bar{\boldsymbol{e}}_1(s_0) \tag{1.2.33}$$

at a fixed parameter value s_0. (Then $\boldsymbol{e}_2(s_0) = \bar{\boldsymbol{e}}_2(s_0)$.) This means that considered as moving points, $\boldsymbol{p}$ and $\bar{\boldsymbol{p}}$ have the same position and velocity at time s_0. Then we have only to prove that in this case the equality $\boldsymbol{p}(s) = \bar{\boldsymbol{p}}(s)$ holds for all s.

First, express components of vectors $\boldsymbol{e}_1, \boldsymbol{e}_2, \bar{\boldsymbol{e}}_1, \bar{\boldsymbol{e}}_2$ as

$$\begin{aligned} \boldsymbol{e}_1 &= (\xi_{11}, \xi_{12}), & \boldsymbol{e}_2 &= (\xi_{21}, \xi_{22}), \\ \bar{\boldsymbol{e}}_1 &= (\bar{\xi}_{11}, \bar{\xi}_{12}), & \bar{\boldsymbol{e}}_2 &= (\bar{\xi}_{21}, \bar{\xi}_{22}), \end{aligned} \tag{1.2.34}$$

and consider two matrices

$$X = \begin{bmatrix} \xi_{11} & \xi_{12} \\ \xi_{21} & \xi_{22} \end{bmatrix}, \qquad \bar{X} = \begin{bmatrix} \bar{\xi}_{11} & \bar{\xi}_{12} \\ \bar{\xi}_{21} & \bar{\xi}_{22} \end{bmatrix}. \tag{1.2.35}$$

Since $\boldsymbol{e}_1$ and $\boldsymbol{e}_2$ are unit vectors which are orthogonal to each other, X is an orthogonal matrix. Similarly, $\bar{X}$ is also an orthogonal matrix. Since $\boldsymbol{p}(s_0) = \bar{\boldsymbol{p}}(s_0)$, in order to prove that $\boldsymbol{p}(s) = \bar{\boldsymbol{p}}(s)$, it suffices to show that $\boldsymbol{p}(s) - \bar{\boldsymbol{p}}(s)$ is a constant vector independent of s by showing that

$$\frac{d}{ds}(\boldsymbol{p}(s) - \bar{\boldsymbol{p}}(s)) = \boldsymbol{0}. \tag{1.2.36}$$

The left hand side of (1.2.36) is equal to $\boldsymbol{e}_1(s) - \bar{\boldsymbol{e}}_1(s)$ so we need to show that $\xi_{11}(s) = \bar{\xi}_{11}(s)$ and $\xi_{12}(s) = \bar{\xi}_{12}(s)$. For this purpose, we prove that

$$\begin{cases} (\xi_{11} - \bar{\xi}_{11})^2 + (\xi_{21} - \bar{\xi}_{21})^2 &= 0, \\ (\xi_{12} - \bar{\xi}_{12})^2 + (\xi_{22} - \bar{\xi}_{22})^2 &= 0. \end{cases} \tag{1.2.37}$$

The important point of this proof is to consider (1.2.37) instead of $(\xi_{11} - \bar{\xi}_{11})^2 + (\xi_{12} - \bar{\xi}_{12})^2 = 0$.

Since X and $\bar{X}$ are orthogonal matrices,

$$\begin{aligned}&(\xi_{11}-\bar{\xi}_{11})^2+(\xi_{21}-\bar{\xi}_{21})^2\\&=\xi_{11}{}^2+\xi_{21}{}^2+\bar{\xi}_{11}{}^2+\bar{\xi}_{21}{}^2-2(\xi_{11}\bar{\xi}_{11}+\xi_{21}\bar{\xi}_{21})\\&=2-2(\xi_{11}\bar{\xi}_{11}+\xi_{21}\bar{\xi}_{21}).\end{aligned} \tag{1.2.38}$$

On the other hand, from $\boldsymbol{e}_1(s_0)=\bar{\boldsymbol{e}}_1(s_0)$ we have $\boldsymbol{e}_2(s_0)=\bar{\boldsymbol{e}}_2(s_0)$. Hence

$$\xi_{11}(s_0)\bar{\xi}_{11}(s_0)+\xi_{21}(s_0)\bar{\xi}_{21}(s_0)=\xi_{11}(s_0)^2+\xi_{21}(s_0)^2=1.$$

Thus, in order to prove $\xi_{11}\bar{\xi}_{11}+\xi_{21}\bar{\xi}_{21}=1$, we differentiate by s and check that $\xi_{11}\bar{\xi}_{11}+\xi_{21}\bar{\xi}_{21}$ is constant. Using (1.2.22) we get

$$\begin{aligned}&\xi_{11}{}'=\kappa\xi_{21}, &&\xi_{12}{}'=\kappa\xi_{22},\\&\bar{\xi}_{11}{}'=\kappa\bar{\xi}_{21}, &&\bar{\xi}_{12}{}'=\kappa\bar{\xi}_{22},\\&\xi_{21}{}'=-\kappa\xi_{11}, &&\xi_{22}{}'=-\kappa\xi_{12},\\&\bar{\xi}_{21}{}'=-\kappa\bar{\xi}_{11}, &&\bar{\xi}_{22}{}'=-\kappa\bar{\xi}_{12},\end{aligned} \tag{1.2.39}$$

and from this,

$$\begin{aligned}&\frac{d}{ds}(\xi_{11}\bar{\xi}_{11}+\xi_{21}\bar{\xi}_{21})\\&=\xi_{11}{}'\bar{\xi}_{11}+\xi_{11}\bar{\xi}_{11}{}'+\xi_{21}{}'\bar{\xi}_{21}+\xi_{21}\bar{\xi}_{21}{}'\\&=\kappa\xi_{21}\bar{\xi}_{11}+\kappa\xi_{11}\bar{\xi}_{21}-\kappa\xi_{11}\bar{\xi}_{21}-\kappa\xi_{21}\bar{\xi}_{11}\\&=0.\end{aligned} \tag{1.2.40}$$

Thus we know that $\xi_{11}\bar{\xi}_{11}+\xi_{21}\bar{\xi}_{21}=1$ and from (1.2.38) we get $(\xi_{11}-\bar{\xi}_{11})^2+(\xi_{21}-\bar{\xi}_{21})^2=0$. In the same way, we obtain $(\xi_{12}-\bar{\xi}_{12})^2+(\xi_{22}-\bar{\xi}_{22})^2=0$ and (1.2.37). From the above proof, we obtain the following theorem.

Theorem 1.2.2 *A necessary and sufficient condition for the curvatures $\kappa(s)$ and $\bar{\kappa}(s)$ of two plane curves $\mathbf{p}(s)$ and $\bar{\mathbf{p}}(s)$ to coincide is that by suitable rotation and parallel displacement $\bar{\mathbf{p}}(s)$ can be superimposed on $\mathbf{p}(s)$.*

Example 1.2.2 Ellipse $x=a\cos t,\ y=b\sin t\ (a>b>0)$.

First, from $\frac{x^2}{a^2}+\frac{y^2}{b^2}=\cos^2 t+\sin^2 t=1$, we know that this equation is an ellipse. So far we have been using the length parameter s in calculating the curvature. However, in the above equation, t is not the length parameter. Changing the parameter t to s is theoretically convenient but at times it is very difficult in practice. As we will see soon, what we need is only $\frac{dt}{ds}$, not $t=t(s)$.

Generally, we have

$$s=\int_0^t\sqrt{\left(\frac{dx}{dt}\right)^2+\left(\frac{dy}{dt}\right)^2}\,dt, \tag{1.2.41}$$

so we get

$$\frac{ds}{dt}=\sqrt{\left(\frac{dx}{dt}\right)^2+\left(\frac{dy}{dt}\right)^2}. \tag{1.2.42}$$

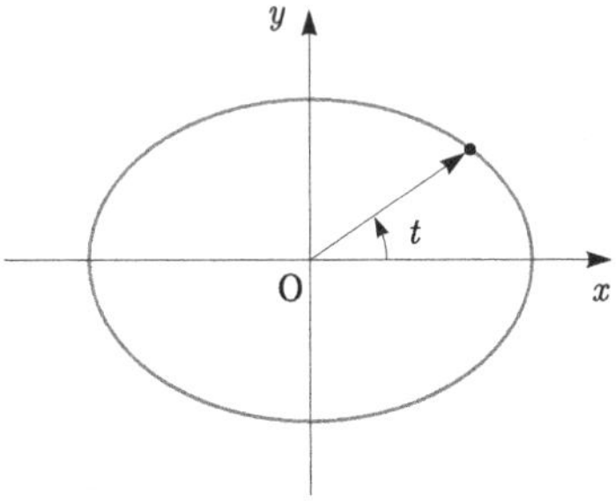

Fig. 1.2.4

For this ellipse we have

$$\frac{ds}{dt} = \sqrt{a^2 \sin^2 t + b^2 \cos^2 t}, \qquad \frac{dt}{ds} = \frac{1}{\sqrt{a^2 \sin^2 t + b^2 \cos^2 t}}. \tag{1.2.43}$$

Next we calculate the vector $\boldsymbol{e}_1$. From

$$\boldsymbol{e}_1 = \frac{d\boldsymbol{p}}{ds} = \frac{d\boldsymbol{p}}{dt}\frac{dt}{ds}, \tag{1.2.44}$$

we have

$$\boldsymbol{e}_1 = \left(\frac{-a \sin t}{\sqrt{a^2 \sin^2 t + b^2 \cos^2 t}}, \frac{b \cos t}{\sqrt{a^2 \sin^2 t + b^2 \cos^2 t}}\right) \tag{1.2.45}$$

and by rotating $\boldsymbol{e}_1$ by 90° in the positive direction we get

$$\boldsymbol{e}_2 = \left(\frac{-b \cos t}{\sqrt{a^2 \sin^2 t + b^2 \cos^2 t}}, \frac{-a \sin t}{\sqrt{a^2 \sin^2 t + b^2 \cos^2 t}}\right). \tag{1.2.46}$$

Using

$$\frac{d\boldsymbol{e}_1}{ds} = \frac{d\boldsymbol{e}_1}{dt}\frac{dt}{ds} \tag{1.2.47}$$

together with (1.2.22), (1.2.45), and (1.2.46) we obtain

$$\kappa(t) = \frac{ab}{(a^2 \sin^2 t + b^2 \cos^2 t)^{\frac{3}{2}}}. \tag{1.2.48}$$

When $a = b$, that is, in the case of a circle, we have $\kappa = \frac{1}{a}$, which agrees with the result of Example 1.2.1. Note that for the above calculation of κ, we need to differentiate only the first component of $\boldsymbol{e}_1$ and compare with the first component of $\boldsymbol{e}_2$. Thus we need not differentiate the second component of $\boldsymbol{e}_1$. ♦

We can generalize the above argument to see that for a curve

$$x = x(t), \qquad y = y(t), \tag{1.2.49}$$

the curvature is given by

$$\kappa(t) = \frac{\dot{x}(t)\ddot{y}(t) - \ddot{x}(t)\dot{y}(t)}{(\dot{x}(t)^2 + \dot{y}(t)^2)^{\frac{3}{2}}}. \tag{1.2.50}$$

This is left to the reader as an exercise.

In order to get a further understanding of this section, it might be useful to review the section on curvatures in your calculus text book. You may also calculate curvatures of curves given in that text book. The following are only review problems on calculus.

Problem 1.2.1 Prove the equation of the curvature (1.2.50).

Problem 1.2.2 When the curve is given as $r = F(\theta)$ in terms of the polar coordinates (r, θ), prove that the curvature is given by

$$\kappa = \frac{r^2 + 2\left(\dfrac{dr}{d\theta}\right)^2 - r\dfrac{d^2r}{d\theta^2}}{\left\{r^2 + \left(\dfrac{dr}{d\theta}\right)^2\right\}^{\frac{3}{2}}}.$$

Problem 1.2.3 Check that a curve

$$x = \cosh t, \quad y = \sinh t$$

becomes a hyperbola

$$x^2 - y^2 = 1,$$

draw its graph, and calculate its curvature. Note that

$$\cosh t = \frac{e^t + e^{-t}}{2}, \quad \sinh t = \frac{e^t - e^{-t}}{2}.$$

Problem 1.2.4 Let a be a positive constant number. Draw the graph of a **catenary**

$$y = a\cosh\frac{x}{a},$$

calculate the arc length s from the point $(0, a)$ to the point $(x, a\cosh(\frac{x}{a}))$, and find the expression of the curve in terms of the parameter s.

1.3 Global Theorems on Plane Curves

In the previous section we studied the curvature $\kappa(s)$ of a curve $\boldsymbol{p}(s)$ and saw that in order to obtain $\kappa(s_0)$ we need to know how $\boldsymbol{p}(s)$ behaves when s moves within a small interval including s_0. Thus we need not worry about $\boldsymbol{p}(s)$ when s is far away from s_0. We obtain $\kappa(s_0)$ from the second derivative of $\boldsymbol{p}(s)$ at s_0, but the derivatives of a function at a particular point is determined by the behavior of the function in a neighborhood of that point. In that way, what we studied in the previous section are **local properties** of a curve. The local theory of plane curves are, in essence, given by the definition of curvature, equation (1.2.23) and Theorem (1.2.2). Whether or not the curve is closed is not a local question.

On the other hand, if the curve is closed, it is compact and there exists a point where the curvature has its maximum and minimum. It is also possible to integrate the curvature on the curve. There arises also problems such as "how many times a closed curve rotates around a particular point." In this section we study such **global properties** about the whole curve.

In order to define the curvature κ, we had to differentiate $\boldsymbol{p}$ twice. Since we need to differentiate κ, from now on we assume that $\boldsymbol{p}$ is differentiable three times. Then, we call a point $\boldsymbol{p}(s_0)$ where $\kappa'(s_0) = 0$ a **vertex** of the curve. It may sound a little strange to call a point of a "smooth" curve a vertex, but from equation (1.2.48), we can see that the curvature of an ellipse in Example 1.2.2 takes the maximum value $\frac{a}{b^2}$ at $t = 0, \pi$ and the minimum value $\frac{b}{a^2}$ at $t = \frac{\pi}{2}, \frac{3\pi}{2}$. Thus κ is maximum at the points of intersection with the x-axis and is minimum at the points of intersection with the y-axis. This agrees with our intuition that the sharper the curve is the larger the curvature. At the point where κ has its maximum or minimum, its derivative κ' is 0. Hence the four points where the ellipse intersects the x- and y- axis are its vertices. Since κ is monotone increasing or decreasing in each of the four parts parted by the vertices, the word "vertex" seems appropriate at least for an ellipse.

A plane curve $\boldsymbol{p}(s)$ $(a \leq s \leq b)$ is called a **closed curve** when the initial point $\boldsymbol{p}(a)$ and its terminal point $\boldsymbol{p}(b)$ are the same. If $\boldsymbol{p}(s_1) \neq \boldsymbol{p}(s_2)$ for other parameter values s_1, s_2 differ from a,b, it is called a **simple closed curve**. The famous Jordan's theorem, which states that every simple closed curve separates the plane into two regions, inner and outer. This topological theorem may sound so obvious and it may be difficult to see if there is anything that has to be proved. Leaving the proof of such a theorem to a professional mathematician, we shall proceed to more concrete problems. Hereafter we assume $\boldsymbol{p}(s)$ is differentiable at least two times and the differentials at the initial point and the terminal point coincides if $\boldsymbol{p}(s)$ is closed. When a line segment connecting arbitrary two points of a simple closed curve does not go outside the region enclosed by the curve, the curve is called an **oval** or a **convex closed curve**.

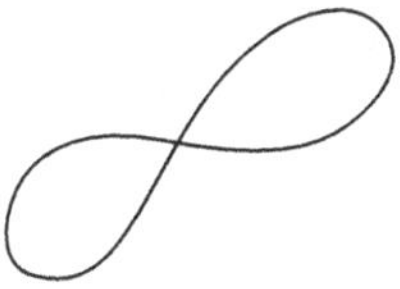

a closed curve, but not a simple closed curve

Fig. 1.3.1

a simple closed curve which is not convex

Fig. 1.3.2

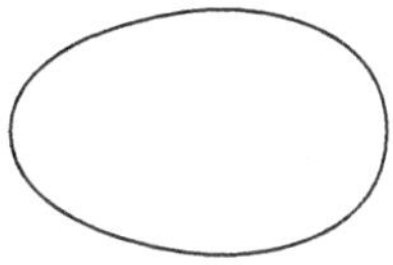

an oval

Fig. 1.3.3

Using what we have explained above, we shall prove the **Four vertex theorem** of Mukhopadhyaya.

Theorem 1.3.1 *Every oval has at least four vertices.*

Proof Let $\boldsymbol{p}(s)$ $(a \leq s \leq b)$ be an oval. Consider the curvature κ as a continuous function defined on a closed interval $[a, b]$. Then as we have learned in calculus, κ has the maximum and minimum. Thus assume that it has its maximum at point M on $\boldsymbol{p}$ and its minimum at point N. Using rotation and parallel displacement, move M and N so that they will be located on the x-axis. Since M and N are both vertices, we need to find two more.

First assume that M and N are the only vertices. Then κ is monotone increasing between N and M, where $\kappa' > 0$, and is monotone decreasing between M and N, where $\kappa' < 0$.

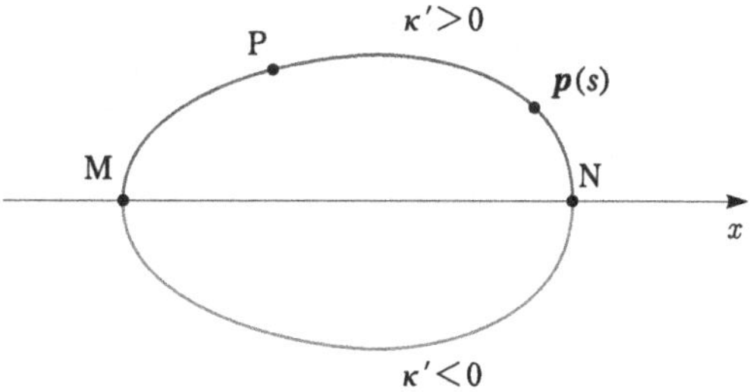

Fig. 1.3.4

Thus, $\kappa'(s)y(s)$ is always positive at points other than M and N, and we get,

$$\int_a^b \kappa'(s)\, y(s)\, ds > 0. \tag{1.3.1}$$

On the other hand, by integrating by parts we have,

$$\begin{aligned}\int_a^b \kappa'(s)\, y(s)\, ds &= [\kappa(s)\, y(s)]_a^b - \int_a^b \kappa(s)\, y'(s)\, ds \\ &= -\int_a^b \kappa(s)\, y'(s)\, ds.\end{aligned} \tag{1.3.2}$$

On the other hand, comparing the second component of $\boldsymbol{e}_2{}' = -\kappa \boldsymbol{e}_1 = -\kappa \boldsymbol{p}'$ of (1.2.22) we get

$$\xi_{22}{}' = -\kappa\, \xi_{12} = -\kappa\, y'. \tag{1.3.3}$$

Thus

$$\int_a^b \kappa'(s)\, y(s)\, ds = \int_a^b \xi_{22}{}'(s)\, ds = [\xi_{22}(s)]_a^b = 0. \tag{1.3.4}$$

This is a contradiction and we know that there exists at least another vertex P, other than M and N.

Let P be located on the arc between N and M. If there are no other vertices other than the three mentioned above, on the arc between M and N we still have $\kappa' < 0$. If there is no point other than P where $\kappa' = 0$ on the arc between N and M, this means that κ does not decrease between N and M. (By drawing a graph of $\kappa(s)$ we can see that if κ decreases somewhere between N and M, we get at least one local maximum and a local minimum in between N and M, where $\kappa' = 0$, thus producing two more vertices). Thus when κ moves from N to M, at points other than P we have $\kappa' > 0$. Thus at points other than M, N, and P, we get $\kappa'(s)y(s) > 0$, and this again contradicts (1.3.4). □

The **four vertex theorem** was proved by S. Mukhopadhyaya from Bengal in 1909 in his paper (New methods in the geometry of a plane arc, Bull. Calcutta Math. Soc. vol. 1 (1909), pp. 31–37). The proof given here is by G. Herglotz, and has become widely known after it had been introduced in the textbook (Vorlesungen über Differentialgeometrie) by W. Blaschke. This theorem has been generalized to all simple closed curve (not necessarily convex), but we will not go into this generalization here.

Next we shall explain what the **rotation index** of a closed curve is. Let κ be the curvature of a closed curve $\boldsymbol{p}(s)$ $(a \le s \le b)$. Then the number m we get from the integration

$$m = \frac{1}{2\pi} \int_a^b \kappa(s)\, ds \tag{1.3.5}$$

is always an integer (i.e., $0, \pm 1, \pm 2, \cdots$). In order to understand why, we have to recall the explanation of the curvature in terms of the Gauss map g. As the parameter s moves from a to b, the point $\boldsymbol{p}(s)$ moves on the curve $\boldsymbol{p}$ with a constant speed 1 from $\boldsymbol{p}(a)$ to $\boldsymbol{p}(b)$. On the other hand, the normal vector $\boldsymbol{e}_2(s)$ moves along a unit circle from $\boldsymbol{e}_2(a)$ to $\boldsymbol{e}_2(b)$ with the speed $\kappa(s)$ then. When $\boldsymbol{e}_2(s)$ moves in the positive direction (anti-clockwise) we have $\kappa(s) > 0$, and when it moves in the negative direction $\kappa(s) < 0$. Thus the distance $\boldsymbol{e}_2(s)$ traveled on the circle, between $\boldsymbol{e}_2(a)$ and $\boldsymbol{e}_2(b)$ is given by,

$$l = \int_a^b \kappa(s)\, ds. \tag{1.3.6}$$

Here we calculate the distance traveled in the negative direction as a negative distance. For example look at the movement of normal vector $e_2(s)$ in Figure 1.3.5. When $p(s)$

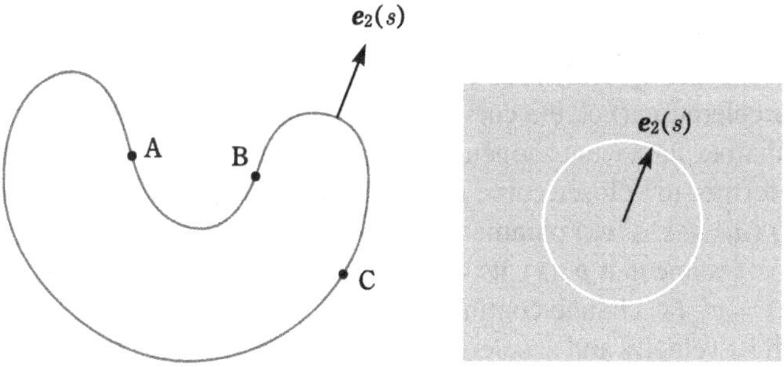

Fig. 1.3.5

moves between points A and B, $e_2(s)$ moves in the positive direction. Between B and C, it moves in the negative direction, and past point C, it faces the positive direction again. (Readers should draw a few normal vectors on the figure and parallelly displace it onto the unit circle to see how $e_2(s)$ moves.)

Since we are assuming the curve is smoothly closed, we have $p(a) = p(b)$ and $e_2(a) = e_2(b)$. On the unit circle, $e_2(s)$ starts from $e_2(a)$ and moves to $e_2(b) = e_2(a)$. It may rotate around the unit circle back and forth but will eventually reach its initial point. Let us call the number of times it really rotated around the circle the **rotation index** of a closed curve p. Check that for Figure 1.3.5, the rotation index is 1. As $e_2(s)$ moves around the unit circle once in the positive direction, it travels the distance of 2π. Hence, the value l we obtain from the integral (1.3.6) will be 2π times the rotation index. Thus, the value m in (1.3.5) is the rotation index and, in particular, it is an integer.

Draw normal vectors $e_2(s)$ for each of the curves in Figures 1.3.6, 1.3.7, and 1.3.8 and check the rotation index by yourself. Do not forget that $e_2(s)$ points to the left of the direction of the motion. Reversing the direction of the movement changes the sign of the rotation index.

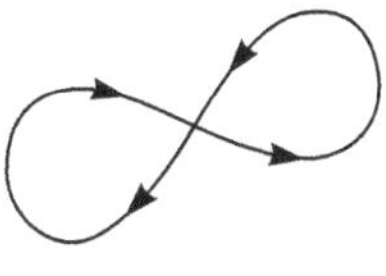

$m=0$

Fig. 1.3.6

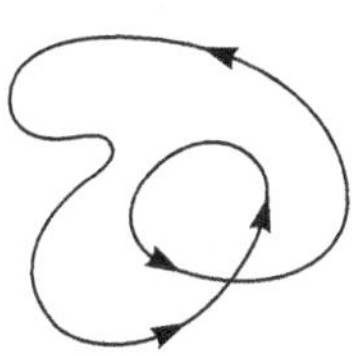

$m=2$

Fig. 1.3.7

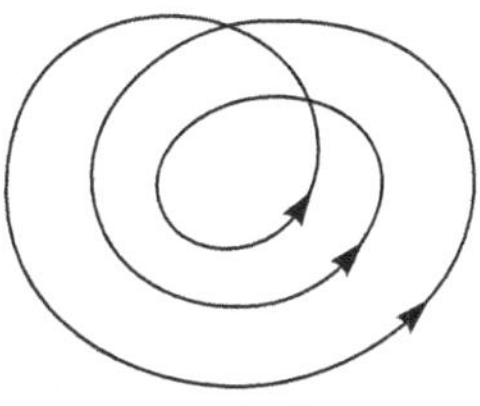

$m=3$

Fig. 1.3.8

We give a physical interpretation of the rotation index and the integral (1.3.5). Assume that you are running along a closed curve p with constant speed without going back and forth. Let the sun be in the east. As you run along the curve, everytime the sun which was on your left moves to your right, count it as +1. Everytime the sun which was on your right moves to your left, count it as −1. If you try this for all of the four figures above you will see that by adding all of the ±1 you get the rotation index. On the other hand, as you are running, if you know the curvature at very instant, by adding them together (integrating them) you get (2π times) the rotation index. This means that theoretically, with your eyes closed you can count the rotation index by feeling the acceleration (i.e., the curvature).

Now we shall explain what happens to the rotation index when closed curve $\boldsymbol{p}(s)$ is gradually deformed to a closed curve $\overline{\boldsymbol{p}}(s)$. More explicitly, we have a family of closed curves $\boldsymbol{p}_r(s)$ ($a_r \leq s \leq b_r$) parametrized by r, $0 \leq r \leq 1$, with $\boldsymbol{p}(s) = \boldsymbol{p}_0(s)$ and $\overline{\boldsymbol{p}}(s) = \boldsymbol{p}_1(s)$. Assume that $\boldsymbol{p}_r(s)$, its velocity vector $\boldsymbol{e}_{1r}(s) = \boldsymbol{p}_r'(s)$, and acceleration vector $\boldsymbol{e}_{2r}(s) = \boldsymbol{e}_{1r}'(s)$ change continuously with r. In other words, we assume that together with its velocity and acceleration, $\overline{\boldsymbol{p}}(s)$ deforms continuously to $\boldsymbol{p}(s)$. Since the curvature $\kappa_r(s)$ of $\boldsymbol{p}_r(s)$ is given by

$$\boldsymbol{e}_{1r}'(s) = \kappa_r(s)\, \boldsymbol{e}_{2r}(s),$$

$\kappa_r(s)$ also changes continuously with r. Therefore, the rotation index

$$m_r = \frac{1}{2\pi}\int_{a_r}^{b_r} \kappa_r(s)\, ds \tag{1.3.7}$$

should also change continuously with r, but since m_r is an integer, it cannot change gradually. Thus m_r does not change at all, meaning that the rotation index of $\boldsymbol{p}(s)$ and the rotation index of $\overline{\boldsymbol{p}}(s)$ are the same. Since the rotation index of the curve in Figure 1.3.5 is 1, and the rotation index of the curve in Figure 1.3.6 is 0, we cannot gradually deform one curve to the other in this case.

Imagine placing a string like the curve in Figure 1.3.6. Then it seems like we can deform it gradually to the curve in Figure 1.3.5, and this may look like a contradiction. However, in order to use the above argument, for every instant the curve is deforming, $\boldsymbol{p}_r(s)$ should be smooth (twice differentiable) and the curvature $\kappa_r(s)$ must be defined. When we are in the process of deforming Figure 1.3.6 to Figure 1.3.5, we get a curve which is not smooth. This notion of a smooth deformation was introduced quite recently, and is often used in differential topology.

In the above, we considered the integration of the curvature $\kappa(s)$. Now we consider the integration of the absolute value of the curvature

$$\mu = \int_a^b |\kappa(s)|\, ds \tag{1.3.8}$$

and call it the **total curvature**. As we did with the rotation index, let us use the Gauss map to think about the total curvature. Integrating $|\kappa(s)|$ instead of $\kappa(s)$ means that no matter which direction the normal vector $\boldsymbol{e}_2(s)$ moves on the unit circle, we calculate the distance it moved as positive. Thus as $\boldsymbol{e}_2(s)$ moves back and forth, the integral increases. Also we can easily check that μ in (1.3.8) is not necessarily an integer multiple of 2π. The next result is a special case of Fenchel's theorem for a curve in three-dimensional space. Since it is very simple we shall prove it here.

Theorem 1.3.2 *The total curvature μ of a closed plane curve $\boldsymbol{p}(s)$ satisfies the inequality $\mu \geq 2\pi$, and the equality $\mu = 2\pi$ holds only when $\boldsymbol{p}(s)$ is an oval.*

First, assume that the range of $\boldsymbol{e}_1(s) = (x'(s), y'(s))$ $(a \leq s \leq b)$, is smaller than a semi-circle, and we shall show that it leads to a contradiction. By rotating the curve, we may assume $\boldsymbol{e}_1(s)$ belongs to the upper half of a circumference. Thus for all s, let $y'(s) > 0$. Therefore we get

$$0 < \int_a^b y'(s)\,ds = y(b) - y(a). \tag{1.3.9}$$

On the other hand, since $\boldsymbol{p}(s)$ is a closed curve $y(b) = y(a)$, and this is a contradiction. Thus $\boldsymbol{e}_1(s)$ $(a \leq s \leq b)$ covers at least half the circumference. In a closed curve, any point can be the initial point, so we can assume that $\boldsymbol{e}_1(s)$ $(a \leq s \leq s_1)$ with $s_1 < b$ covers exactly half the circumference and we have $\boldsymbol{e}_1(a) = -\boldsymbol{e}_1(s_1)$. Since, we get $\boldsymbol{e}_2(s)$ by rotating $\boldsymbol{e}_1(s)$ by 90°, we have $\boldsymbol{e}_2(a) = -\boldsymbol{e}_2(s_1)$. Therefore when s moves from a to s_1, and also when it moves from s_1 to b, the distance $\boldsymbol{e}_2(s)$ travels is at least π. Thus,

$$\int_a^{s_1} |\kappa(s)|\,ds \geq \pi, \quad \int_{s_1}^b |\kappa(s)|\,ds \geq \pi. \tag{1.3.10}$$

By combining the two inequalities we get $\mu \geq 2\pi$. Now let $\mu = 2\pi$. Then $\kappa(s)$ is always $\kappa(s) \geq 0$ or always $\kappa(s) \leq 0$ and $\boldsymbol{e}_2(s)$ rotates along the unit circle once, without going back and forth. When $\kappa(s) \leq 0$, by changing the direction of the curve $\boldsymbol{p}(s)$, we get $\kappa(s) \geq 0$. Thus we continue our proof assuming that $\mu = 2\pi$ and $\kappa(s) \geq 0$.

In order to prove that the curve $\boldsymbol{p}(s)$ is a simple closed curve, assume that there exists an s_0 such that $\boldsymbol{p}(a) = \boldsymbol{p}(s_0)$ and $a < s_0 < b$. Since we can rotate the curve, we may assume that neither $\boldsymbol{e}_1(a)$ nor $\boldsymbol{e}_1(s_0)$ is horizontal. Separate the curve $\boldsymbol{p}(s) = (x(s), y(s))$ into two parts $a \leq s \leq s_0$ and $s_0 \leq s \leq b$. If necessary, by rotating the curve again we may assume $y(a) = y(s_0) = y(b)$ is neither the minimum nor the maximum of $y(s)$ on both closed intervals $[a, s_0]$ and $[s_0, b]$. Then in each part the maximum value of $y(s)$ is larger than $y(a) = y(s_0)$, and the minimum value of $y(s)$ is smaller than $y(a) = y(s_0)$. At the maximum and minimum points of $y(s)$, the normal vector $\boldsymbol{e}_2(s)$ is in the vertical direction. Thus, there are at least four points where $\boldsymbol{e}_2(s)$ is vertical, and this contradicts the fact that when s changes from a to b, $\boldsymbol{e}_2(s)$ moves exactly once along the unit circle without going back and forth (see Figure 1.3.9).

Now we know that $\boldsymbol{p}(s)$ is a simple closed curve. In order to prove that it is convex, we need to prove that the curve $\boldsymbol{p}(s)$ lies on one side of its tangent line at each point. Draw a tangent at any point $\boldsymbol{p}(s_0)$ on the curve. Since we can rotate the curve, assume that the tangent is horizontal. If the curve $\boldsymbol{p}(s) = (x(s), y(s))$ is on both sides of the horizontal tangent at $\boldsymbol{p}(s_0)$, there exists a maximum point $y(s_1)$ and minimum point $y(s_2)$ for $y(s)$. The tangent vectors $\boldsymbol{e}_1(s_0)$, $\boldsymbol{e}_1(s_1)$, and $\boldsymbol{e}_1(s_2)$ at $\boldsymbol{p}(s_0)$, $\boldsymbol{p}(s_1)$,

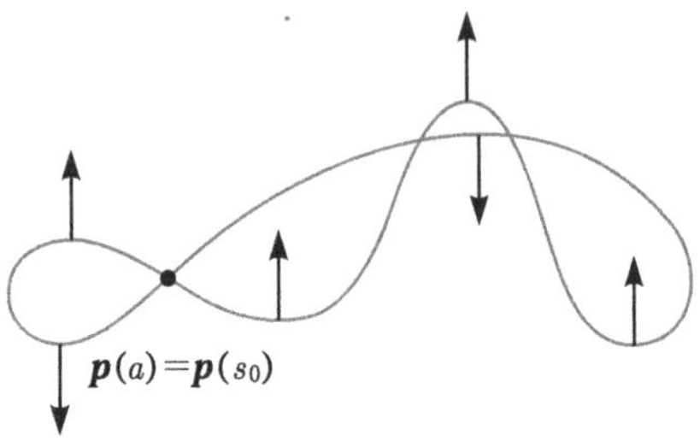

Fig. 1.3.9

and $p(s_2)$ are horizontal. Thus normal vectors $e_2(s_0)$, $e_2(s_1)$, and $e_2(s_2)$ are vertical. This contradicts the fact that $e_2(s)$ goes around the unit circle exactly once without going back and forth. Hence, $p(s)$ is on one side of the tangent line at $p(s_0)$. Thus we know that $p(s)$ is an oval.

In studying properties of an oval, we can use another convenient parameter beside the length parameter s. Thus for each t $(0 \leq t \leq 2\pi)$, there exists exactly one point on the oval where the unit normal vector e_2 is $(\cos t, \sin t)$. In other words, the Gauss map g is one-to-one. So we can write s as a function of t, $s = s(t)$, and t as a function of s, $t = t(s)$; as t moves from 0 to 2π, s moves from a to b. From the definition of the parameter t (see Figure 1.3.10),

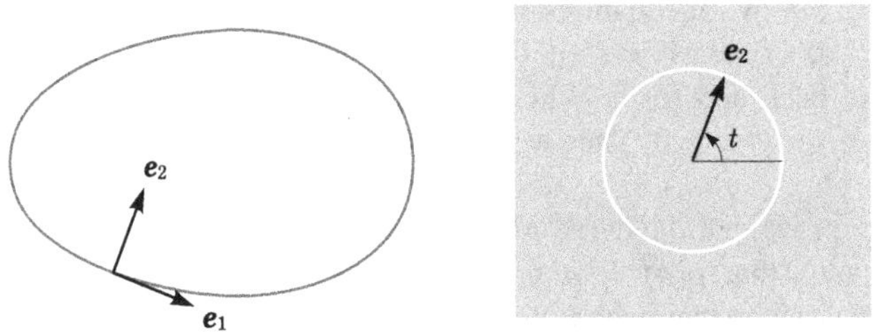

Fig. 1.3.10

$$e_1(s(t)) = (\sin t, -\cos t), \quad e_2(s(t)) = (\cos t, \sin t). \tag{1.3.11}$$

Thus, from (1.2.22) and (1.3.11)

$$\kappa\, e_2 = \frac{de_1}{ds} = \frac{de_1}{dt}\frac{dt}{ds} = (\cos t, \sin t)\frac{dt}{ds} = e_2\frac{dt}{ds}, \tag{1.3.12}$$

and we get

$$\kappa = \frac{dt}{ds}. \tag{1.3.13}$$

(By this we merely reconfirmed the relationship between the Gauss map and the curvature we proved in the previous section.)

For simplicity's sake, let us write $\boldsymbol{p}(t)$ instead of $\boldsymbol{p}(s(t))$. Using this parameter t, the **width** of the oval $\boldsymbol{p}(t)$ can be defined easily. Now for one fixed t_0, the tangents at $\boldsymbol{p}(t_0)$ and $\boldsymbol{p}(t_0+\pi)$ are parallel to each other pointing in the opposite direction. We call this width between the two tangents the width of $\boldsymbol{p}(t)$ at t_0 and write it as $W(t_0)$ (see Figure 1.3.11).

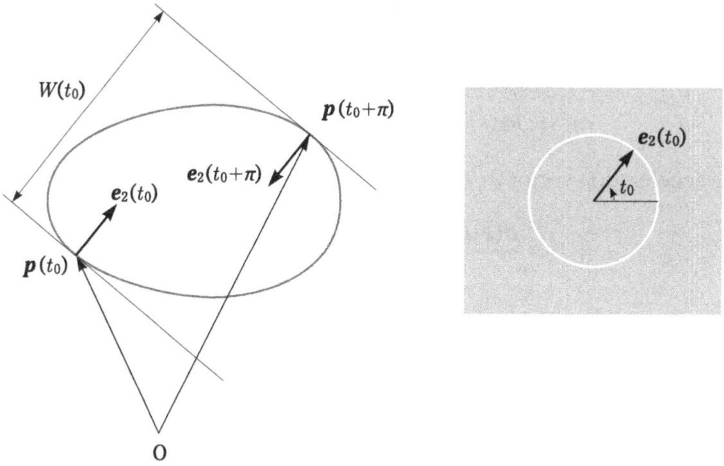

Fig. 1.3.11

Problem 1.3.1 (a) Check that the width $W(t)$ defined above is given by

$$W(t) = -\boldsymbol{p}(t) \cdot \boldsymbol{e}_2(t) - \boldsymbol{p}(t+\pi) \cdot \boldsymbol{e}_2(t+\pi).$$

(b) Prove that the length L of an oval $\boldsymbol{p}(t)$ is given by

$$L = \int_0^{\pi} W(t)\, dt.$$

When the width W is constant, we get

$$L = \pi W.$$

(This is called Barbier's theorem and was proved in 1806.)

The following problem is for readers who have learned complex analysis.

Problem 1.3.2 Let $w = f(z)$ be a complex analytic function defined in a region including the unit circle $|z| \leq 1$. Assume the zeros of $f'(z)$ are not located on the circle $|z| = 1$. When z moves along the circle $|z| = 1$, calculate the curvature and rotation index of the curve described by w.

1.4 Space Curves

Let $\boldsymbol{p}(s) = (x(s), y(s), z(s))$ $(a \leq s \leq b)$ be a curve in the three-dimensional Euclidean space. As in Section 1.2, let s be a parameter such that the velocity of the movement given by the curve $\boldsymbol{p}(s)$ is a constant equal to 1. Thus, the velocity vector

$$\boldsymbol{e}_1(s) = \boldsymbol{p}'(s) = (x'(s), y'(s), z'(s)) \tag{1.4.1}$$

has length 1. More explicitly, we have

$$\boldsymbol{e}_1(s) \cdot \boldsymbol{e}_1(s) = x'(s)^2 + y'(s)^2 + z'(s)^2 = 1. \tag{1.4.2}$$

Then the acceleration vector $\boldsymbol{e}_1{}'(s)$ satisfies

$$0 = \frac{d(\boldsymbol{e}_1(s) \cdot \boldsymbol{e}_1(s))}{ds} = 2\boldsymbol{e}_1{}'(s) \cdot \boldsymbol{e}_1(s),$$

so $\boldsymbol{e}_1{}'(s)$ is perpendicular to $\boldsymbol{e}_1(s)$. Write the length of $\boldsymbol{e}_1{}'(s)$ as $\kappa(s)$ and call it the **curvature** of the curve $\boldsymbol{p}(s)$. Thus,

$$\begin{aligned}\kappa(s) &= \sqrt{\boldsymbol{e}_1{}'(s) \cdot \boldsymbol{e}_1{}'(s)} \\ &= \sqrt{x''(s)^2 + y''(s)^2 + z''(s)^2}.\end{aligned} \tag{1.4.3}$$

From the definition of $\kappa(s)$, we have $\kappa(s) \geq 0$. When we studied plane curves, we defined $\kappa(s)$ by $\boldsymbol{e}_1{}'(s) = \kappa(s)\,\boldsymbol{e}_2(s)$ where $\boldsymbol{e}_2(s)$ is the vector $\boldsymbol{e}_1(s)$ rotated by 90°. Thus we have

$$|\kappa(s)| = \sqrt{\boldsymbol{e}_1{}'(s) \cdot \boldsymbol{e}_1{}'(s)},$$

but for a plane curve $\kappa(s)$ was negative at times. However for space curves, there are infinite number of unit vectors perpendicular to $\boldsymbol{e}_1(s)$, so we cannot define $\boldsymbol{e}_2(s)$ simply from $\boldsymbol{e}_1(s)$. When the curvature defined in (1.4.3) is not zero, we can define $\boldsymbol{e}_2(s)$ by the equation

$$\boldsymbol{e}_1{}'(s) = \kappa(s)\,\boldsymbol{e}_2(s). \tag{1.4.4}$$

Thus we may define $\boldsymbol{e}_2(s)$ as

$$\boldsymbol{e}_2(s) = \frac{1}{\kappa(s)}\,\boldsymbol{e}_1{}'(s).$$

Therefore $\boldsymbol{e}_2(s)$ is a unit vector which has the same direction as the acceleration vector $\boldsymbol{e}_1{}'(s)$. However when $\kappa(s) = 0$, $\boldsymbol{e}_2(s)$ cannot be uniquely defined.

From now on we shall continue with the assumption $\kappa(s) \neq 0$, so let $\kappa(s) > 0$.

When $\boldsymbol{e}_1(s)$ and $\boldsymbol{e}_2(s)$ are fixed, we can set $\boldsymbol{e}_3(s)$ as a unit vector perpendicular to both $\boldsymbol{e}_1(s)$ and $\boldsymbol{e}_2(s)$. More precisely, since there are two possibilities for choosing $\boldsymbol{e}_3(s)$, we determine $\boldsymbol{e}_3(s)$ by the condition that $\boldsymbol{e}_1(s)$, $\boldsymbol{e}_2(s)$, and $\boldsymbol{e}_3(s)$ are a right-handed system, as in Figure 1.4.1. Hence by rotation and parallel displacement, $\boldsymbol{e}_1(s)$, $\boldsymbol{e}_2(s)$, and $\boldsymbol{e}_3(s)$ can be moved to the positive direction of the x-axis, y-axis,

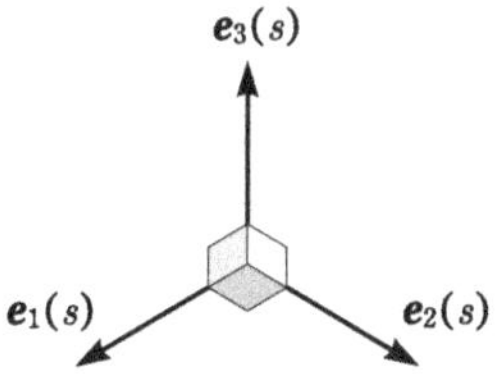

Fig. 1.4.1

and z-axis respectively. Using the vector product $\times$ we have $\boldsymbol{e}_3 = \boldsymbol{e}_1 \times \boldsymbol{e}_2$. We call these $\boldsymbol{e}_1$, $\boldsymbol{e}_2$, $\boldsymbol{e}_3$ the **Frenet frame**. We call the line obtained by extending $\boldsymbol{e}_2(s)$ the **principal normal**, and the line obtained by extending $\boldsymbol{e}_3(s)$ the **binormal**. By differentiating

$$\boldsymbol{e}_i \cdot \boldsymbol{e}_j = \delta_{ij} \quad (\delta_{ij} = 1 \text{ when } i = j, \text{ and } \delta_{ij} = 0 \text{ when } i \neq j), \tag{1.4.5}$$

we get

$$\boldsymbol{e}_i{}' \cdot \boldsymbol{e}_j + \boldsymbol{e}_i \cdot \boldsymbol{e}_j{}' = 0. \tag{1.4.6}$$

In particular, when $i = 1$, $j = 2$,

$$0 = \boldsymbol{e}_1{}' \cdot \boldsymbol{e}_2 + \boldsymbol{e}_1 \cdot \boldsymbol{e}_2{}' = \kappa\, \boldsymbol{e}_2 \cdot \boldsymbol{e}_2 + \boldsymbol{e}_1 \cdot \boldsymbol{e}_2{}' = \kappa + \boldsymbol{e}_1 \cdot \boldsymbol{e}_2{}'. \tag{1.4.7}$$

When $i = j = 2$, we have

$$0 = 2\, \boldsymbol{e}_2 \cdot \boldsymbol{e}_2{}'. \tag{1.4.8}$$

From (1.4.8), we know that $\boldsymbol{e}_2{}'$ is perpendicular to $\boldsymbol{e}_2$, thus it is a linear combination of $\boldsymbol{e}_1$ and $\boldsymbol{e}_3$ and from (1.4.7), the coefficient of $\boldsymbol{e}_1$ in the linear combination is $-\kappa$. By writing the coefficient of $\boldsymbol{e}_3$ as τ, we have

$$\boldsymbol{e}_2{}' = -\kappa \boldsymbol{e}_1 + \tau \boldsymbol{e}_3. \tag{1.4.9}$$

For (1.4.6), when $i = 1$ and $j = 3$,

$$0 = \boldsymbol{e}_1{}' \cdot \boldsymbol{e}_3 + \boldsymbol{e}_1 \cdot \boldsymbol{e}_3{}' = \kappa\, \boldsymbol{e}_2 \cdot \boldsymbol{e}_3 + \boldsymbol{e}_1 \cdot \boldsymbol{e}_3{}' = \boldsymbol{e}_1 \cdot \boldsymbol{e}_3{}'. \tag{1.4.10}$$

When $i = 2$ and $j = 3$,

$$\begin{aligned} 0 &= \boldsymbol{e}_2{}' \cdot \boldsymbol{e}_3 + \boldsymbol{e}_2 \cdot \boldsymbol{e}_3{}' \\ &= (-\kappa\, \boldsymbol{e}_1 + \tau\, \boldsymbol{e}_3) \cdot \boldsymbol{e}_3 + \boldsymbol{e}_2 \cdot \boldsymbol{e}_3{}' = \tau + \boldsymbol{e}_2 \cdot \boldsymbol{e}_3{}'. \end{aligned} \tag{1.4.11}$$

When $i = j = 3$, we get

$$0 = 2\, \boldsymbol{e}_3{}' \cdot \boldsymbol{e}_3, \tag{1.4.12}$$

and from (1.4.10), (1.4.11), and (1.4.12) we get

$$\boldsymbol{e}_3{}' = -\tau\, \boldsymbol{e}_2. \tag{1.4.13}$$

From the above we have

$$\begin{cases} p' = e_1, \\ e_1{}' = \qquad \kappa\, e_2, \\ e_2{}' = -\kappa\, e_1 \qquad + \tau\, e_3, \\ e_3{}' = \qquad -\tau\, e_2, \end{cases} \tag{1.4.14}$$

thus

$$\begin{bmatrix} e_1{}' \\ e_2{}' \\ e_3{}' \end{bmatrix} = \begin{bmatrix} 0 & \kappa & 0 \\ -\kappa & 0 & \tau \\ 0 & -\tau & 0 \end{bmatrix} \begin{bmatrix} e_1 \\ e_2 \\ e_3 \end{bmatrix}$$

and call (1.4.14) **Frenet-Serret's formula** of a space curve, and τ the **torsion**.

Let a curve $p(s)$ be on the plane, for example $a \cdot p = c$ (a is a vector perpendicular to the plane, c is a constant). Since

$$a \cdot p(s) = c, \tag{1.4.15}$$

by differentiating it and using (1.4.14), we have

$$a \cdot e_1 = 0, \qquad a \cdot \kappa\, e_2 = 0. \tag{1.4.16}$$

Then using $\kappa \neq 0$, we obtain $a \cdot e_2 = 0$, and a is perpendicular to e_1 and e_2. Also by differentiating $a \cdot e_2 = 0$ and from (1.4.14) we have

$$a \cdot (-\kappa\, e_1 + \tau\, e_3) = 0. \tag{1.4.17}$$

Therefore $a \cdot \tau e_3 = 0$. Since a is already perpendicular to e_1 and e_2, it is not perpendicular to e_3, so $\tau = 0$.

Conversely, when the torsion τ of a curve $p(s)$ is zero, from (1.4.14) we have $e_3{}' = 0$. Thus e_3 is a constant vector independent of s. Since the derivative of $e_3 \cdot p(s)$ is equal to $e_3 \cdot e_1 = 0$, $e_3 \cdot p(s)$ is a constant c not depending on s. Hence, $p(s)$ lies on the plane $e_3 \cdot p = c$. Summarizing, we have the following theorem.

Theorem 1.4.1 *Assume that the curvature κ of a space curve $p(s)$ is everywhere positive. Then the torsion τ is zero everywhere if and only if $p(s)$ lies on the plane.*

Let us calculate the curvature κ and the torsion τ of simple examples.

Example 1.4.1 A **helix** is given by

$$x = a \cos t, \quad y = a \sin t, \quad z = bt \quad (a > 0). \tag{1.4.18}$$

First, we calculate the velocity vectors

$$\dot{x} = -a \sin t, \quad \dot{y} = a \cos t, \quad \dot{z} = b, \tag{1.4.19}$$

and the length (speed) is

$$\sqrt{\dot{x}^2 + \dot{y}^2 + \dot{z}^2} = \sqrt{a^2 \sin^2 t + a^2 \cos^2 t + b^2} = \sqrt{a^2 + b^2}. \tag{1.4.20}$$

Thus in order for the speed to be 1, the parameter s must be given by

$$s = \sqrt{a^2 + b^2}\, t. \tag{1.4.21}$$

By rewriting (1.4.18) we have

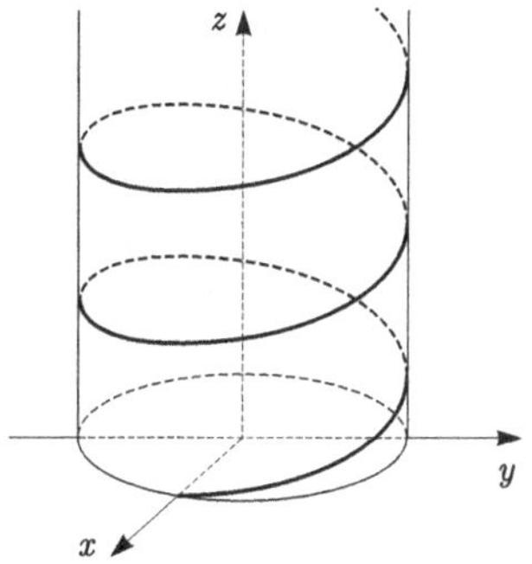

Fig. 1.4.2

$$x = a\cos\frac{s}{c}, \quad y = a\sin\frac{s}{c}, \quad z = \frac{b}{c}s \qquad (c = \sqrt{a^2 + b^2}). \tag{1.4.22}$$

Thus by differentiating (1.4.22) and using (1.4.14) we have

$$\boldsymbol{e}_1 = (x', y', z') = \left(-\frac{a}{c}\sin\frac{s}{c}, \frac{a}{c}\cos\frac{s}{c}, \frac{b}{c}\right), \tag{1.4.23}$$

$$\boldsymbol{e}_1{}' = \left(-\frac{a}{c^2}\cos\frac{s}{c}, -\frac{a}{c^2}\sin\frac{s}{c}, 0\right), \tag{1.4.24}$$

$$\kappa = \sqrt{\boldsymbol{e}_1{}' \cdot \boldsymbol{e}_1{}'} = \frac{a}{c^2}, \tag{1.4.25}$$

$$\boldsymbol{e}_2 = \frac{1}{\kappa}\boldsymbol{e}_1{}' = \left(-\cos\frac{s}{c}, -\sin\frac{s}{c}, 0\right), \tag{1.4.26}$$

$$\boldsymbol{e}_3 = \boldsymbol{e}_1 \times \boldsymbol{e}_2 = \left(\frac{b}{c}\sin\frac{s}{c}, -\frac{b}{c}\cos\frac{s}{c}, \frac{a}{c}\right), \tag{1.4.27}$$

$$\boldsymbol{e}_2{}' = \left(\frac{1}{c}\sin\frac{s}{c}, -\frac{1}{c}\cos\frac{s}{c}, 0\right) = -\frac{a}{c^2}\boldsymbol{e}_1 + \frac{b}{c^2}\boldsymbol{e}_3, \tag{1.4.28}$$

and so

$$\kappa = \frac{a}{a^2 + b^2}, \quad \tau = \frac{b}{a^2 + b^2}. \tag{1.4.29}$$

Thus the curvature κ and the torsion τ are both constants. ◆

In general, given a space curve with both $\kappa(> 0)$ and τ constant, by rotation and parallel displacement we move it to the helix of the form (1.4.18). In that case, the constants a, b are determined from (1.4.29) by κ and τ. This can be proved directly from Frenet-Serret's formula, but can also be obtained as a special case of the following general theorem on uniqueness.

Theorem 1.4.2 *Let the curvature and the torsion of two space curves* $\boldsymbol{p}(s)$ *and* $\overline{\boldsymbol{p}}(s)$ *be* $\kappa(s)$, $\tau(s)$, $\overline{\kappa}(s)$, *and* $\overline{\tau}(s)$, *respectively. Then a necessary and sufficient condition for* $\kappa = \overline{\kappa}$ *and* $\tau = \overline{\tau}$ *is that by suitable rotation and parallel displacement* $\bar{\boldsymbol{p}}(s)$ *can be superimposed on* $\boldsymbol{p}(s)$.

Proof The proof of this theorem is almost the same as the proof of Theorem 1.2.2, so we will discuss only the changes we have to make. As in (1.2.33), let

$$\begin{aligned} \boldsymbol{p}(s_0) &= \bar{\boldsymbol{p}}(s_0), & \boldsymbol{e}_1(s_0) &= \bar{\boldsymbol{e}}_1(s_0), \\ \boldsymbol{e}_2(s_0) &= \bar{\boldsymbol{e}}_2(s_0), & \boldsymbol{e}_3(s_0) &= \bar{\boldsymbol{e}}_3(s_0). \end{aligned} \tag{1.4.30}$$

As in (1.2.34) we have

$$\begin{aligned} \boldsymbol{e}_1 &= (\xi_{11}, \xi_{12}, \xi_{13}), & \boldsymbol{e}_2 &= (\xi_{21}, \xi_{22}, \xi_{23}), & \boldsymbol{e}_3 &= (\xi_{31}, \xi_{32}, \xi_{33}), \\ \bar{\boldsymbol{e}}_1 &= (\bar{\xi}_{11}, \bar{\xi}_{12}, \bar{\xi}_{13}), & \bar{\boldsymbol{e}}_2 &= (\bar{\xi}_{21}, \bar{\xi}_{22}, \bar{\xi}_{23}), & \bar{\boldsymbol{e}}_3 &= (\bar{\xi}_{31}, \bar{\xi}_{32}, \bar{\xi}_{33}), \end{aligned} \tag{1.4.31}$$

and orthogonal matrices

$$X = \begin{bmatrix} \xi_{11} & \xi_{12} & \xi_{13} \\ \xi_{21} & \xi_{22} & \xi_{23} \\ \xi_{31} & \xi_{32} & \xi_{33} \end{bmatrix}, \quad \bar{X} = \begin{bmatrix} \bar{\xi}_{11} & \bar{\xi}_{12} & \bar{\xi}_{13} \\ \bar{\xi}_{21} & \bar{\xi}_{22} & \bar{\xi}_{23} \\ \bar{\xi}_{31} & \bar{\xi}_{32} & \bar{\xi}_{33} \end{bmatrix}. \tag{1.4.32}$$

Since $\boldsymbol{p}(s_0) = \bar{\boldsymbol{p}}(s_0)$, in order to show $\boldsymbol{p}(s) = \bar{\boldsymbol{p}}(s)$, we have only to prove

$$\frac{d}{ds}(\boldsymbol{p}(s) - \bar{\boldsymbol{p}}(s)) = \boldsymbol{0}. \tag{1.4.33}$$

However, because the left-hand side of (1.4.33) is $\boldsymbol{e}_1(s) - \bar{\boldsymbol{e}}_1(s)$, this is reduced to proving

$$\xi_{11}(s) = \bar{\xi}_{11}(s), \quad \xi_{12}(s) = \bar{\xi}_{12}(s), \quad \xi_{13}(s) = \bar{\xi}_{13}(s).$$

For this, it suffices to prove

$$\begin{cases} (\xi_{11} - \bar{\xi}_{11})^2 + (\xi_{21} - \bar{\xi}_{21})^2 + (\xi_{31} - \bar{\xi}_{31})^2 &= 0, \\ (\xi_{12} - \bar{\xi}_{12})^2 + (\xi_{22} - \bar{\xi}_{22})^2 + (\xi_{32} - \bar{\xi}_{32})^2 &= 0, \\ (\xi_{13} - \bar{\xi}_{13})^2 + (\xi_{23} - \bar{\xi}_{23})^2 + (\xi_{33} - \bar{\xi}_{33})^2 &= 0. \end{cases} \tag{1.4.34}$$

To prove the first equation of (1.4.34), write it as

$$\begin{aligned} &(\xi_{11} - \bar{\xi}_{11})^2 + (\xi_{21} - \bar{\xi}_{21})^2 + (\xi_{31} - \bar{\xi}_{31})^2 \\ &\qquad = 2 - 2(\xi_{11}\bar{\xi}_{11} + \xi_{21}\bar{\xi}_{21} + \xi_{31}\bar{\xi}_{31}). \end{aligned} \tag{1.4.35}$$

Since $\xi_{11}\bar{\xi}_{11} + \xi_{21}\bar{\xi}_{21} + \xi_{31}\bar{\xi}_{31} = 1$ at $s = s_0$, we have only to prove that this is a constant number (independent of s). Writing out Frenet-Serret's formula by vector components, we get

$$\begin{cases} \xi_{11}' = \kappa\,\xi_{21}, & \xi_{12}' = \kappa\,\xi_{22}, & \xi_{13}' = \kappa\,\xi_{23}, \\ \xi_{21}' = -\kappa\,\xi_{11} + \tau\,\xi_{31}, & \xi_{22}' = -\kappa\,\xi_{12} + \tau\,\xi_{32}, & \xi_{23}' = -\kappa\,\xi_{13} + \tau\,\xi_{33}, \\ \xi_{31}' = -\tau\,\xi_{21}, & \xi_{32}' = -\tau\,\xi_{22}, & \xi_{33}' = -\tau\,\xi_{23}. \end{cases} \tag{1.4.36}$$

We get the same equation for $\bar{\xi}_{ij}$. Using (1.4.36) and the barred equation (1.4.36) we can prove

$$\frac{d}{ds}(\xi_{11}\bar{\xi}_{11} + \xi_{21}\bar{\xi}_{21} + \xi_{31}\bar{\xi}_{31}) = 0. \tag{1.4.37}$$

The proofs for the remaining equations of (1.4.34) are similar. □

In Section 1.2 we defined the Gauss map for a plane curve and studied its general properties in Section 1.3. We can also define **Gauss' spherical map** for a space curve $\boldsymbol{p} = \boldsymbol{p}(s)$. We get the map g from a curve $\boldsymbol{p}$ into the unit sphere by parallel displacing the velocity vector $\boldsymbol{e}_1(s_0)$ to the origin of the space (see Figure 1.4.3).

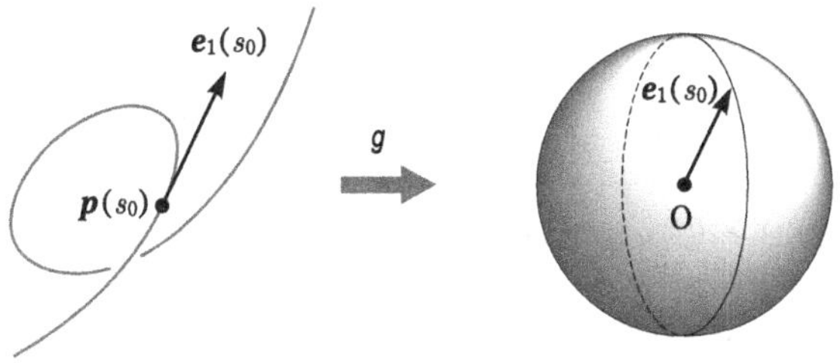

Fig. 1.4.3

In order to define the Gauss map for a plane curve, we used in Section 1.2 the normal vector $\boldsymbol{e}_2(s_0)$. In that case $\boldsymbol{e}_1(s_0)$ and $\boldsymbol{e}_2(s_0)$ differed by a rotation by 90°. Thus it did not matter whether we used $\boldsymbol{e}_1(s_0)$ or $\boldsymbol{e}_2(s_0)$. However, for a space curve, in order to use $\boldsymbol{e}_2(s)$, κ has to be positive everywhere. So using $\boldsymbol{e}_1(s)$ is more convenient when defining the spherical map for a space curve. What $\boldsymbol{e}_1' = \kappa\boldsymbol{e}_2$ in Frenet-Serret's formula (1.4.14) means is that when $\boldsymbol{p}(s)$ moves with constant speed 1 in the space, $\boldsymbol{e}_1(s)$ moves on the unit sphere with speed κ. In case of plane curves, according as $\boldsymbol{e}_2(s)$ moves in the positive or negative direction, κ becomes positive or negative. In case of space curves, it is meaningless to talk about whether $\boldsymbol{e}_1(s)$ moves in the positive direction or negative direction on the unit sphere. Note that by definition, κ is always positive or zero.

In Theorem 1.4.2 we proved that a space curve is determined by its curvature κ and torsion τ. It is also important to prove that given functions $\kappa(s) > 0$ and $\tau(s)$ of s, there exists a space curve which has κ as its curvature and τ as its torsion. However, in order to prove this assertion, we need to know about the existence theorem for ordinary differential equations (see the explanation of differential equations at the end of the book). Since we do not use this result later, those who do not know the existence theorem for ordinary differential equations can just glance through this explanation. Given functions $\kappa(s) > 0$ and $\tau(s)$, let $\boldsymbol{e}_1$, $\boldsymbol{e}_2$, $\boldsymbol{e}_3$ be unknown vector functions of s, and consider a system of ordinary differential equations

$$\begin{cases} {e_1}' = & \kappa\, e_2, \\ {e_2}' = -\kappa\, e_1 & +\tau\, e_3, \\ {e_3}' = & -\tau\, e_2. \end{cases} \tag{1.4.38}$$

As its initial condition, for example, if $\boldsymbol{e}_1(0)$, $\boldsymbol{e}_2(0)$, and $\boldsymbol{e}_3(0)$ are given as a right-handed frame, from the existence and uniqueness of a solution for ordinary differential equations, we see that there exists one and only one solution $\boldsymbol{e}_1(s)$, $\boldsymbol{e}_2(s)$, and $\boldsymbol{e}_3(s)$ for (1.4.38). The curve $\boldsymbol{p}(s)$ we are seeking is given by

$$\boldsymbol{p}(s) = \int_0^s \boldsymbol{e}_1(s)\, ds + \boldsymbol{c}, \tag{1.4.39}$$

if we want $\boldsymbol{c}$ to be its initial point. By calculation, we can see that the curve defined by (1.4.39) has the curvature κ and torsion τ.

To familiarize the reader with calculations on space curves, we give three problems and end this section.

Problem 1.4.1 Prove that the Taylor expansion of a space curve $\boldsymbol{p}(s)$ at $s = 0$ to its third term is given by

$$\begin{aligned} \boldsymbol{p}(s) &= \boldsymbol{p}(0) + \boldsymbol{e}_1(0)s + \kappa(0)\boldsymbol{e}_2(0)\frac{s^2}{2} \\ &\quad + \{-\kappa(0)^2\boldsymbol{e}_1(0) + \kappa'(0)\boldsymbol{e}_2(0) + \kappa(0)\tau(0)\boldsymbol{e}_3(0)\}\frac{s^3}{3!} + \cdots . \end{aligned} \tag{1.4.40}$$

This is called **Bouquet's formula** and is helpful in seeing the effect κ and τ have on the shape of the curve.

Problem 1.4.2 Sometimes, space curves are defined by a general parameter t rather than the length parameter s. In that case, it is more convenient to calculate the curvature and torsion directly, without changing the parameter to s. For a space curve $\boldsymbol{p}(t)$ prove

$$\kappa = \frac{|\dot{\boldsymbol{p}} \times \ddot{\boldsymbol{p}}|}{|\dot{\boldsymbol{p}}|^3}, \qquad \tau = \frac{|\dot{\boldsymbol{p}}\ \ddot{\boldsymbol{p}}\ \dddot{\boldsymbol{p}}|}{|\dot{\boldsymbol{p}} \times \ddot{\boldsymbol{p}}|^2}. \tag{1.4.41}$$

Here $\times$ is the vector product, $|\ \ |$ is the length of a vector, and $|\dot{\boldsymbol{p}}\ \ddot{\boldsymbol{p}}\ \dddot{\boldsymbol{p}}|$ is the determinant of the matrix of size three made from the components of the three vectors $\dot{\boldsymbol{p}}$, $\ddot{\boldsymbol{p}}$, and $\dddot{\boldsymbol{p}}$.

The first equation of (1.4.41) is the generalization of (1.2.50) from plane curves to space curves.

Problem 1.4.3 Let $\boldsymbol{p}(s) = (x^1(s), x^2(s), \cdots, x^n(s))$ be a curve in the n-dimensional Euclidean space, which is parameterized in such a way that its speed is 1. Then, prove that the Frenet-Serret's formula is

$$\begin{aligned} &{\boldsymbol{e}_i}' = -\kappa_{i-1}\, \boldsymbol{e}_{i-1} + \kappa_i\, \boldsymbol{e}_{i+1} && (i = 1, 2, \cdots, n), \\ &\kappa_0 = \kappa_n = 0, \quad \kappa_j > 0 && (j = 1, 2, \cdots, n-2). \end{aligned} \tag{1.4.42}$$

Just as we assumed that the length κ of $e_1{}'$ is positive in the three-dimensional case, for n-dimensions we need to make suitable assumption. It is important to state the assumption clearly when proving the formula.

1.5 Global Results on Space Curves

In order to define the torsion τ of a space curve, we had to define a principal normal e_2, and to do so we needed the assumption that the curvature κ is positive everywhere. Thus, when a curve is given, its torsion is not necessarily defined. Therefore, when talking about global properties of a curve, it is usually impossible to discuss things involving the torsion, and we will be restricted to discussing properties relating to the curvature.

In Section 1.3, we proved Fenchel's theorem for plane curves, but this theorem was originally proved for space curves.

Theorem 1.5.1 (Fenchel's theorem) *Let $\kappa(s)$ be the curvature of a closed curve $p(s)$ $(a \leq s \leq b)$ in the space. Then*

$$\int_a^b \kappa(s)ds \geq 2\pi \tag{1.5.1}$$

and the equality holds only when $p(s)$ is an oval lying on the plane.

We call the left-hand side of (1.5.1) the **total curvature** of a closed curve. As a preparation for proving Theorem 1.5.1 we will prove the formula for the total curvature when the curve is expressed as $q(t)$ $(a \leq t \leq b)$ by any parameter t. Consider the unit vector

$$e = \frac{\dot{q}}{(\dot{q} \cdot \dot{q})^{\frac{1}{2}}} \tag{1.5.2}$$

obtained by dividing the velocity vector $\dot{q}(t)$ by its length. Using the parameter s, the curvature κ_q of q is $\kappa_q = \left(\frac{de}{ds} \cdot \frac{de}{ds}\right)^{\frac{1}{2}}$ so note that the total curvature is (letting $\alpha \leq s \leq \beta$ for $a \leq t \leq b$),

$$\int_\alpha^\beta \left(\frac{de}{ds} \cdot \frac{de}{ds}\right)^{\frac{1}{2}} ds = \int_\alpha^\beta \left(\frac{de}{dt} \cdot \frac{de}{dt}\right)^{\frac{1}{2}} \frac{dt}{ds} ds = \int_a^b (\dot{e} \cdot \dot{e})^{\frac{1}{2}} dt. \tag{1.5.3}$$

By differentiating (1.5.2) we get

$$\dot{e} = \frac{(\dot{q} \cdot \dot{q})\ddot{q} - (\dot{q} \cdot \ddot{q})\dot{q}}{(\dot{q} \cdot \dot{q})^{\frac{3}{2}}}. \tag{1.5.4}$$

Therefore,

$$\dot{e} \cdot \dot{e} = \frac{(\dot{q} \cdot \dot{q})(\ddot{q} \cdot \ddot{q}) - (\dot{q} \cdot \ddot{q})^2}{(\dot{q} \cdot \dot{q})^2}. \tag{1.5.5}$$

Substituting this into (1.5.3) we get the total curvature

$$\int_a^b \frac{\{(\dot{q} \cdot \dot{q})(\ddot{q} \cdot \ddot{q}) - (\dot{q} \cdot \ddot{q})^2\}^{\frac{1}{2}}}{\dot{q} \cdot \dot{q}} \, dt. \tag{1.5.6}$$

The formula (1.5.6) can be used for both space curves and plane curves.

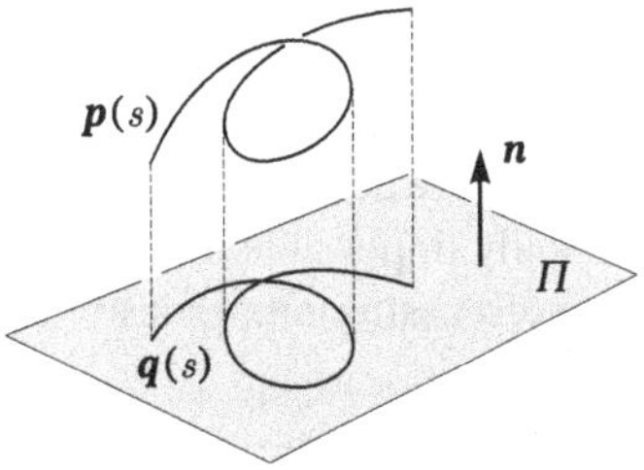

Fig. 1.5.1

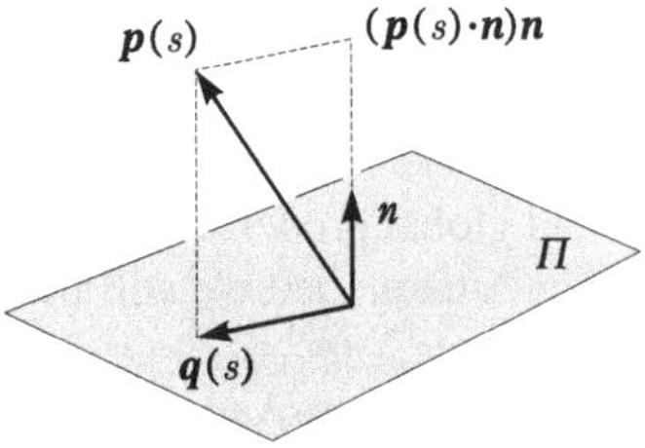

Fig. 1.5.2

Next, take a unit vector $\boldsymbol{n}$ which is located at the origin of the space and let Π be the plane orthogonal to $\boldsymbol{n}$ at the origin. Let $\boldsymbol{q}(s)$ be the plane curve obtained by projecting a space curve $\boldsymbol{p}(s)$ onto Π (see Figure 1.5.1). Then, $\boldsymbol{q}(s)$ is

$$\boldsymbol{q}(s) = \boldsymbol{p}(s) - (\boldsymbol{p}(s) \cdot \boldsymbol{n})\boldsymbol{n}, \tag{1.5.7}$$

which is obvious from Figure 1.5.2. From the definition of the parameter s, $\boldsymbol{p}'(s)$ is a unit vector but $\frac{d\boldsymbol{q}}{ds}$ is not necessarily a unit vector. Thus we write $\dot{\boldsymbol{q}}$ for $\frac{d\boldsymbol{q}}{ds}$. Differentiating (1.5.7) by s we get

$$\dot{\boldsymbol{q}} = \boldsymbol{p}' - (\boldsymbol{p}' \cdot \boldsymbol{n})\boldsymbol{n}, \quad \ddot{\boldsymbol{q}} = \boldsymbol{p}'' - (\boldsymbol{p}'' \cdot \boldsymbol{n})\boldsymbol{n}. \tag{1.5.8}$$

We substitute this into (1.5.6) and calculate the total curvature of the plane curve $\boldsymbol{q}(s)$. Since $\boldsymbol{p}' \cdot \boldsymbol{p}' = 1$, using $\boldsymbol{p}' \cdot \boldsymbol{p}'' = 0$ we have

$$\begin{cases} \dot{\boldsymbol{q}} \cdot \dot{\boldsymbol{q}} &= 1 - (\boldsymbol{p}' \cdot \boldsymbol{n})^2, \\ \dot{\boldsymbol{q}} \cdot \ddot{\boldsymbol{q}} &= -(\boldsymbol{p}' \cdot \boldsymbol{n})(\boldsymbol{p}'' \cdot \boldsymbol{n}), \\ \ddot{\boldsymbol{q}} \cdot \ddot{\boldsymbol{q}} &= \boldsymbol{p}'' \cdot \boldsymbol{p}'' - (\boldsymbol{p}'' \cdot \boldsymbol{n})^2. \end{cases} \tag{1.5.9}$$

Thus, by simple calculation, we can write the total curvature $\mu(\boldsymbol{n})$ of $\boldsymbol{q}(s)$ as

$$\mu(\boldsymbol{n}) = \int_a^b \frac{\{\boldsymbol{p}'' \cdot \boldsymbol{p}'' - (\boldsymbol{p}'' \cdot \boldsymbol{n})^2 - (\boldsymbol{p}' \cdot \boldsymbol{n})^2(\boldsymbol{p}'' \cdot \boldsymbol{p}'')\}^{\frac{1}{2}}}{1 - (\boldsymbol{p}' \cdot \boldsymbol{n})^2} \, ds. \tag{1.5.10}$$

Next, since $\boldsymbol{p}' = \boldsymbol{e}_1$ and $\boldsymbol{p}'' = \boldsymbol{e}_1' = \kappa \boldsymbol{e}_2$ by Frenet-Serret's formula, we can rewrite (1.5.10) as follows.

$$\mu(\boldsymbol{n}) = \int_a^b \frac{\kappa\{1-(\boldsymbol{e}_1\cdot\boldsymbol{n})^2-(\boldsymbol{e}_2\cdot\boldsymbol{n})^2\}^{\frac{1}{2}}}{1-(\boldsymbol{e}_1\cdot\boldsymbol{n})^2}\,ds. \tag{1.5.11}$$

In the above calculation we defined $\boldsymbol{e}_2$ by $\boldsymbol{p}'' = \kappa\boldsymbol{e}_2$ by implicitly assuming that $\kappa \neq 0$. However at a point where $\kappa = 0$, for any choice of $\boldsymbol{e}_2$ orthogonal to $\boldsymbol{e}_1$, the integrand in (1.5.11) is zero so there is no problem. The denominator $1-(\boldsymbol{e}_1\cdot\boldsymbol{n})^2$ of (1.5.11) is equal to $\dot{\boldsymbol{q}}\cdot\dot{\boldsymbol{q}}$ by (1.5.9). However, $\dot{\boldsymbol{q}}(s) \neq 0$ may not hold everywhere. For example, when the space curve defined by

$$x = \cos^3 t, \quad y = \sin^3 t, \quad z = \sin\left(t + \frac{\pi}{4}\right) \quad (0 \leq t \leq 2\pi) \tag{1.5.12}$$

is projected onto the (x, y)-plane, we have $\dot{\boldsymbol{q}}(t) = \boldsymbol{0}$ at $t = 0, \frac{\pi}{2}, \pi, \frac{3\pi}{2}$. The graph of $x = \cos^3 t$, $y = \sin^3 t$ has cusps at $t = 0, \frac{\pi}{2}, \pi, \frac{3\pi}{2}$ as in Figure 1.5.3. For a curve of this kind we can also define its total curvature, and will discuss it later. But from what we have studied so far (1.5.11) is true only when $\dot{\boldsymbol{q}} \neq \boldsymbol{0}$ everywhere. Thus, we have to investigate for which unit vector $\boldsymbol{n}$ we have $\dot{\boldsymbol{q}} \neq \boldsymbol{0}$.

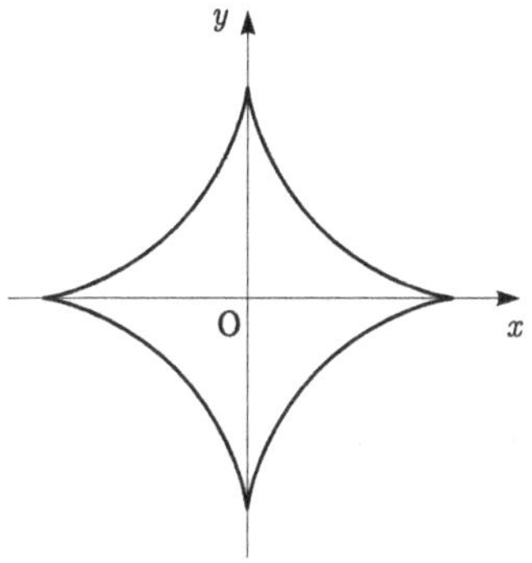

Fig. 1.5.3

Since $\dot{\boldsymbol{q}}(s_0)$ is the projection of $\boldsymbol{p}'(s_0)$ onto the plane Π, we get $\dot{\boldsymbol{q}}(s_0) = \boldsymbol{0}$ exactly when $\boldsymbol{p}'(s_0)$ is orthogonal to Π, that is, parallel to $\boldsymbol{n}$. Since $\boldsymbol{p}'(s_0)$ is a unit vector, considered as a vector at the origin, it is equal to $\pm\boldsymbol{n}$. So the condition $\dot{\boldsymbol{q}}(s_0) = \boldsymbol{0}$ amounts to $\boldsymbol{p}'(s_0) = \pm\boldsymbol{n}$. Since $\boldsymbol{p}'(s)$ $(a \leq s \leq b)$ is a curve on the unit sphere, we get $\dot{\boldsymbol{q}}(s) \neq \boldsymbol{0}$ everywhere if we choose $\boldsymbol{n}$ so that neither $\boldsymbol{n}$ nor $-\boldsymbol{n}$ is on the curve. When a smooth curve $\boldsymbol{p}'(s)$ $(a \leq s \leq b)$ is on a unit sphere which is a two-dimensional surface, most unit vectors $\boldsymbol{n}$ have this property. Namely, the set $\{\pm\boldsymbol{p}'(s); a \leq s \leq b\}$ is a very small set on the sphere and we can choose $\boldsymbol{n}$ from its complement. (The reader who knows measure theory may call it a set with measure zero rather than "a very small set".) With such a choice of $\boldsymbol{n}$ we have $\dot{\boldsymbol{q}}(s) \neq \boldsymbol{0}$ for all s.

The next step in the proof of Theorem 1.5.1 is to prove that considering $\mu(\boldsymbol{n})$ as a function on a unit sphere and averaging it out on the unit sphere we get the total curvature of the original curve $\boldsymbol{p}(s)$. "Averaging" indicates integrating $\mu(\boldsymbol{n})$ on the sphere and dividing it by 4π (the surface area of the unit sphere). When integrating

$\mu(\boldsymbol{n})$, we can exclude a set such as a curve which has area zero (more precisely, measure zero). So we can ignore the set of $\boldsymbol{n}$ such that $\mu(\boldsymbol{n})$ is not defined. What we want to prove is

$$\frac{1}{4\pi}\int_{S^2}\mu(\boldsymbol{n})\,d\boldsymbol{n} = \int_a^b \kappa\,ds. \tag{1.5.13}$$

Here S^2 is the two-dimensional unit sphere and $d\boldsymbol{n}$ is the area element of the sphere S^2 (we will explicitly write it out later). Rewriting (1.5.13) using (1.5.11) we get

$$\begin{aligned}\frac{1}{4\pi}\int_{S^2}\mu(\boldsymbol{n})\,d\boldsymbol{n} &= \frac{1}{4\pi}\int_{S^2}\int_a^b \frac{\kappa\{1-(\boldsymbol{e}_1\cdot\boldsymbol{n})^2-(\boldsymbol{e}_2\cdot\boldsymbol{n})^2\}^{\frac{1}{2}}}{1-(\boldsymbol{e}_1\cdot\boldsymbol{n})^2}\,ds\,d\boldsymbol{n}\\ &= \frac{1}{4\pi}\int_a^b \kappa\int_{S^2}\frac{\{1-(\boldsymbol{e}_1\cdot\boldsymbol{n})^2-(\boldsymbol{e}_2\cdot\boldsymbol{n})^2\}^{\frac{1}{2}}}{1-(\boldsymbol{e}_1\cdot\boldsymbol{n})^2}\,d\boldsymbol{n}\,ds.\end{aligned} \tag{1.5.14}$$

Thus to prove (1.5.13) we have only to check that

$$\int_{S^2}\frac{\{1-(\boldsymbol{e}_1\cdot\boldsymbol{n})^2-(\boldsymbol{e}_2\cdot\boldsymbol{n})^2\}^{\frac{1}{2}}}{1-(\boldsymbol{e}_1\cdot\boldsymbol{n})^2}\,d\boldsymbol{n} = 4\pi. \tag{1.5.15}$$

This equation is for a fixed s and the integral on the left has nothing to do with the curve $\boldsymbol{p}(s)$. Thus let $\boldsymbol{e}_1, \boldsymbol{e}_2, \boldsymbol{e}_3$ be vectors with components $(1, 0, 0)$, $(0, 1, 0)$, $(0, 0, 1)$. Let the component of $\boldsymbol{n}$ be (u, v, w). Then using $u^2 + v^2 + w^2 = 1$,

$$\frac{\{1-(\boldsymbol{e}_1\cdot\boldsymbol{n})^2-(\boldsymbol{e}_2\cdot\boldsymbol{n})^2\}^{\frac{1}{2}}}{1-(\boldsymbol{e}_1\cdot\boldsymbol{n})^2} = \frac{(1-u^2-v^2)^{\frac{1}{2}}}{1-u^2} = \frac{|w|}{v^2+w^2}. \tag{1.5.16}$$

To integrate the function of (1.5.16) on S^2, we divide S^2 into two hemispheres $u \geq 0$ and $u \leq 0$, integrate it on one of the hemispheres, and then multiply the result by two. The hemisphere $u \geq 0$ is defined by

$$u = \sqrt{1-v^2-w^2}. \tag{1.5.17}$$

Generally, in order to calculate the area of a surface defined by

$$u = f(v, w), \tag{1.5.18}$$

as we learned in calculus, we should integrate

$$\iint \sqrt{1+\left(\frac{\partial f}{\partial v}\right)^2+\left(\frac{\partial f}{\partial w}\right)^2}\,dv\,dw \tag{1.5.19}$$

in a region on the (v, w)-plane. In this case, from (1.5.17) and (1.5.19) we have

$$\iint_D \frac{dv\,dw}{\sqrt{1-v^2-w^2}} \tag{1.5.20}$$

and we should integrate on the region $D = \{(v, w); v^2 + w^2 \leq 1\}$ surrounded by a unit circle. Thus the area element $d\boldsymbol{n}$ of S^2 is given as

$$d\boldsymbol{n} = \frac{dv\,dw}{\sqrt{1 - v^2 - w^2}}. \tag{1.5.21}$$

From (1.5.15), (1.5.16), and (1.5.21) we know that we only have to prove

$$2\iint_D \frac{|w|\,dv\,dw}{(v^2 + w^2)\sqrt{1 - v^2 - w^2}} = 4\pi. \tag{1.5.22}$$

In order to integrate this, we may divide D into two parts $w \geq 0$ and $w \leq 0$ and integrate it on $w \geq 0$. Then multiply the result by two. Using the polar coordinates r and θ with $v = r\cos\theta$, $w = r\sin\theta$, we rewrite (1.5.22) to obtain

$$4\int_0^{\pi} d\theta \int_0^1 \frac{\sin\theta\,dr}{\sqrt{1 - r^2}} = 4\,[-\cos\theta]_0^{\pi}\,\left[\sin^{-1} r\right]_0^1 = 4\pi, \tag{1.5.23}$$

thus proving (1.5.22).

We will prove Fenchel's theorem using (1.5.13). In Section 1.3 we proved that the total curvature of a closed curve on a plane is at least 2π. Thus $\mu(\boldsymbol{n}) \geq 2\pi$, and since (1.5.13) is the average of $\mu(\boldsymbol{n})$, its left-hand side is also at least 2π. Next, assume that the total curvature of $\boldsymbol{p}(s)$ is 2π. Since $\mu(\boldsymbol{n}) \geq 2\pi$ and its average is 2π, $\mu(\boldsymbol{n}) = 2\pi$. (From (1.5.11) note that $\mu(\boldsymbol{n})$ is a continuous function of $\boldsymbol{n}$ at places other than $\{\pm\boldsymbol{p}'(s); a \leq s \leq b\}$. Thus, $\mu(\boldsymbol{n}) = 2\pi$ holds for all $\boldsymbol{n}$ which do not belong in $\{\pm\boldsymbol{p}'(s); a \leq s \leq b\}$.) In Section 1.3 we proved that when a closed curve on the plane has total curvature 2π, it is an oval. Thus by projecting $\boldsymbol{p}(s)$ along vector $\boldsymbol{n}$, we obtain a curve $\boldsymbol{q}(s)$ on plane Π which is convex. This is true for every direction $\boldsymbol{n}$. (When $\boldsymbol{n}$ belongs to the set $\{\pm\boldsymbol{p}'(s); a \leq s \leq b\}$, choose a sequence of unit vectors $\boldsymbol{n}_i$ which do not belong to $\{\pm\boldsymbol{p}'(s); a \leq s \leq b\}$ and converges to $\boldsymbol{n}$. Then project $\boldsymbol{p}(s)$ along $\boldsymbol{n}_i$ onto the plane Π_i orthogonal to $\boldsymbol{n}_i$. Then the resulting curve $\boldsymbol{q}_i(s)$ on Π_i converges to $\boldsymbol{q}(s)$. As the limit of a sequence of convex closed curves $\boldsymbol{q}_i(s)$, $\boldsymbol{q}(s)$ is also a convex closed curve.)

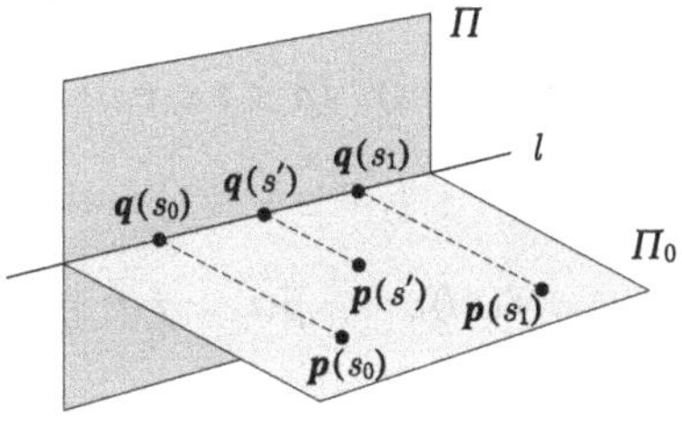

Fig. 1.5.4

Next, to prove that $\boldsymbol{p}(s)$ is a plane curve, assume that it is not and consider an interval $s_0 \leq s \leq s_1$ in which it is not a plane curve. Since $\boldsymbol{p}(s_0)$, $\boldsymbol{p}(s')$, and $\boldsymbol{p}(s_1)$

are not on the same line for a suitably chosen s' ($s_0 \le s' \le s_1$), in this interval, there exists a unique plane Π_0 which goes through these three points. Then choose a plane Π which passes through the origin and is orthogonal to Π_0, and call their line of intersection l. By orthogonally projecting $\boldsymbol{p}(s)$ onto Π, we obtain a convex closed curve $\boldsymbol{q}(s)$ on Π, but $\boldsymbol{q}(s_0)$, $\boldsymbol{q}(s')$, and $\boldsymbol{q}(s_1)$ are on l. For a suitable choice of Π, $\boldsymbol{q}(s_0)$, $\boldsymbol{q}(s')$, and $\boldsymbol{q}(s_1)$ will be three distinct points. Since $\boldsymbol{q}(s)$ is convex, $\boldsymbol{q}(s)$, $s_0 \le s \le s_1$, should be on l. Thus $\boldsymbol{p}(s)$, $s_0 \le s \le s_1$, should be on Π_0 and this is a contradiction. This completes the proof of Fenchel's theorem.

This theorem was first proved by Fenchel in 1929, and since then H. Liebmann, B. Segre, K. Borsuk, H. Rutishauser, H. Samelson and a few others have given different proofs. Interested readers may read *On the differential geometry of closed space curves, Bulletin Amer. Math. Soc. vol. 57 (1951), pp. 45–54* by Fenchel. The proof by Rutishauser-Samelson is quite easy, and it can be found in the paper by S. S. Chern *Curves and surfaces in Euclidean space* in Studies in Math., vol. 4, Studies in Global Geometry and Analysis published by the Math. Assoc. of America.

The proof we mentioned above is based on the proof by Fary, and can be used for the next theorem which says that the total curvature of a knot is at least 4π.

Let $\boldsymbol{p}(t)$ and $\overline{\boldsymbol{p}}(t)$ $(a \le t \le b)$ be two simple closed curves in the space. (Since we are not using the length parameter s, we can use the same interval $a \le t \le b$ for both $\boldsymbol{p}(t)$ and $\overline{\boldsymbol{p}}(t)$.) If we are able to gradually deform the curve $\boldsymbol{p}(t)$ to $\overline{\boldsymbol{p}}(t)$ keeping it a simple closed curve during the deformation, we say that these two curves are **isotopic** to each other.

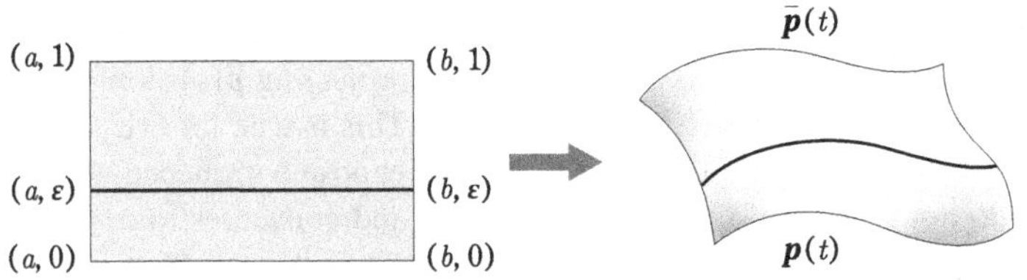

Fig. 1.5.5

More formally stated, this means that there exists a vector valued function

$$\boldsymbol{f}(t,\varepsilon) = (x(t,\varepsilon),\ y(t,\varepsilon),\ z(t,\varepsilon)) \quad (a \le t \le b,\ 0 \le \varepsilon \le 1) \tag{1.5.24}$$

defined on a rectangle $\{(t,\varepsilon);\ a \le t \le b,\ 0 \le \varepsilon \le 1\}$ such that

$$\boldsymbol{p}(t) = \boldsymbol{f}(t,0), \qquad \overline{\boldsymbol{p}}(t) = \boldsymbol{f}(t,1), \tag{1.5.25}$$

and that the curve defined by

$$\boldsymbol{p}_\varepsilon(t) = \boldsymbol{f}(t,\varepsilon) \tag{1.5.26}$$

is a simple closed curve for every ε.

If the curve $\boldsymbol{p}(t)$ is made of a rope and if we are able to gradually move the rope to the position of $\overline{\boldsymbol{p}}(t)$ (stretching and shrinking in the process if necessary), we can say that $\boldsymbol{p}(t)$ and $\overline{\boldsymbol{p}}(t)$ are isotopic. If a rope tied in the form of a ring moves without being untied or cut, then the rope remains simple during the process. A simple closed curve is called a **knot**, and if it is not isotopic to a circle, it is called a **non-trivial** knot. Figure 1.5.7 can deformed into Figure 1.5.6, so it is trivial, but Figure 1.5.8 cannot be deformed into Figure 1.5.6, so it is a non-trivial knot. The definition of a knot mentioned above is an example of how mathematicians try to describe the physical situation mathematically, and there always is a doubt as to whether the mathematical definition accurately reflects the reality. Actually, the definition of a knot mathematicians use is somewhat different from the one given here.

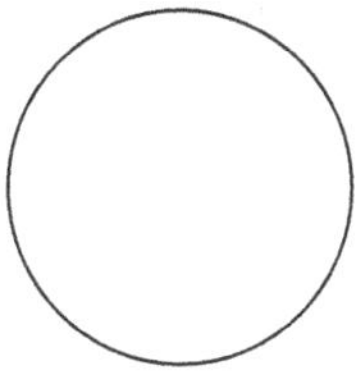

Fig. 1.5.6

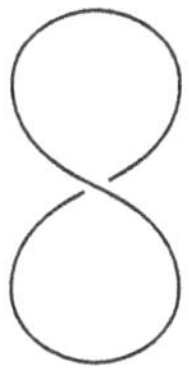

Fig. 1.5.7

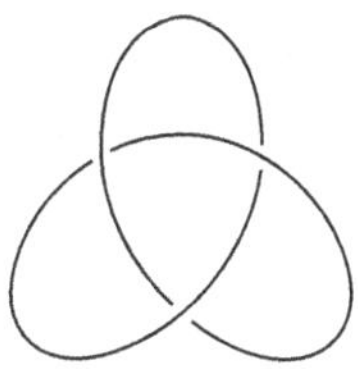

Fig. 1.5.8

Those who want to study knot theory may read *Knotentheorie* (1932) by K. Reidemeister or *Introduction to knot theory* (1963) by R. H. Crowell and R. H. Fox. The next theorem was conjectured by Borsuk in 1948, and was proved by Fary and Milnor independently. (I. Fary, *Sur la courbure totale d'une courbe gauche faisant un noeud*, Bulletin Soc. Math. France. vol. 77 (1949), pp. 128–138; J. W. Milnor, *On the total curvature of knots*, Annals of Math. vol. 52 (1950), pp. 248–257.)

Theorem 1.5.2 *If a simple closed curve $\boldsymbol{p}(s)$ in the space is a non-trivial knot, its total curvature is at least* 4π.

As in the proof of Theorem 1.5.1, when the plane curve $\boldsymbol{q}(s)$ obtained by projecting $\boldsymbol{p}(s)$ in a direction $\boldsymbol{n}$ has total curvature $\mu(\boldsymbol{n})$, the total curvature of $\boldsymbol{p}(s)$ is the average of $\mu(\boldsymbol{n})$. Thus we have to prove that $\mu(\boldsymbol{n}) \geq 4\pi$. (We consider, of course, only those directions $\boldsymbol{n}$ such that $\dot{\boldsymbol{q}}(s) \neq \boldsymbol{0}$.)

Assuming that for some $\boldsymbol{n}$, $\mu(\boldsymbol{n}) < 4\pi$, we will show a contradiction. On a plane Π perpendicular to $\boldsymbol{n}$, consider the Gauss map we used in Section 1.2 for the curve $\boldsymbol{q}(s)$ on Π. Fix a unit circle on Π and let its center be the origin. For each unit vector $\boldsymbol{e}$ on Π, let us count the number of points on the curve $\boldsymbol{q}(s)$ which corresponds to $\boldsymbol{e}$ or $-\boldsymbol{e}$ under the Gauss map. Thus we count the points where the normal vector to $\boldsymbol{q}(s)$ is parallel to $\boldsymbol{e}$. If this number is always at least m for all $\boldsymbol{e}$, then $\mu(\boldsymbol{n}) \geq m\pi$. Therefore, for some $\boldsymbol{e}$, there are at most three such points.

Consider a point of the curve $\boldsymbol{q}(s)$ where the normal vector is parallel to $\boldsymbol{e}$, i.e., a point where the tangent is perpendicular to $\boldsymbol{e}$. In other words, this is a point where the first derivative of the function $\boldsymbol{q}(s) \cdot \boldsymbol{e}$ is zero, and no more than three such points exist. Since $\boldsymbol{q}(s)$ is a closed curve, the number of local maxima and the number of local minima must be the same for the function $\boldsymbol{q}(s) \cdot \boldsymbol{e}$. Thus, there is one local maximum and one local minimum, each. Therefore, if we set the coordinate axis in the space so that the z-axis is in the direction of $\boldsymbol{e}$ and the x-axis is in the direction of

$\boldsymbol{n}$, we get $\boldsymbol{p}(s) = (x(s), y(s), z(s))$, $\boldsymbol{q}(s) = (0, y(s), z(s))$, and there are only one local minimum and one local maximum for $z(s)$.

Set a parameter s in such way that $z(a) = z(b)$ is the local minimum. Then $z(s)$ monotone increasing till one point, then monotone decreasing from there to $z(b)$. Let the maximum be $z(s_0)$, then for each number h between $z(a)$ and $z(s_0)$ there are exactly two points with height h on the curve $\boldsymbol{p}(s)$, i.e., points where $z(s) = h$. We connect these two points with a horizontal line. Then we get a surface which fills the space between curve $\boldsymbol{p}(s)$ (or a film spanning the curve $\boldsymbol{p}(s)$). This surface has a one-to-one, continuous correspondence with a disk $\{(0, y, z); y^2 + z^2 \leq 1\}$. In such a situation (i.e., when there is a film spanning the simple closed curve and there is such a correspondence between the film and the disk), we may define curve $\boldsymbol{p}(s)$ to be a trivial knot. In this sense, we have proved Theorem 1.5.2.

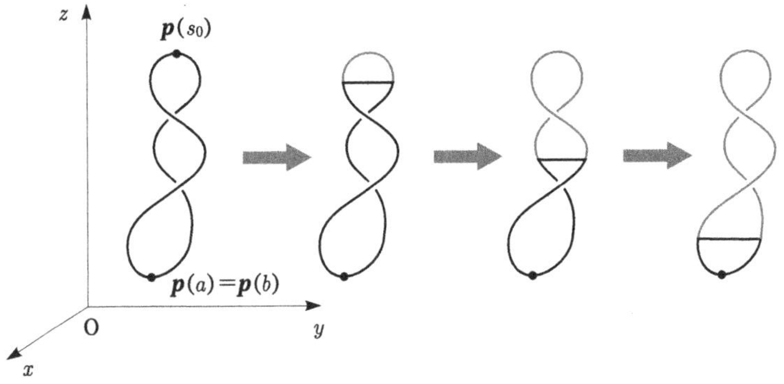

Fig. 1.5.9

Hence, in order to prove that $\boldsymbol{p}(s)$ is isotopic to a circle according to the above definition, join two points on $\boldsymbol{p}(s)$ with the same height by a horizontal line and replace by this line segment the part of the curve $\boldsymbol{p}(s)$ above the line, then gradually lower the height of the line connecting these points, see Figure 1.5.9. During this process, the curve remains simple closed. As in the last figure, when the line is low enough, you can see that $\boldsymbol{p}(s)$ is isotopic to a circle. The above figure is very simple, but even for a complicated curve, if the line is sufficiently lowered, as in Figure 1.5.9, it will become simple enough to be obviously isotopic to a circle.

The above proof is close to that of Fary, but both Fary and Milnor defined the total curvature for simple closed polygons first, and used it in an essential way. A proof using the "Crofton formula" from integral geometry can be found in the survey paper of Fenchel mentioned after the proof of Theorem 1.5.1. It is also in Chern's paper, and is worth reading. In another paper of Milnor *On the total curvatures of closed space curves*, Mathematica Scandinavia, vol. 1 (1953) pp. 289–296, global results involving both curvatures and torsions are proven. There are many papers about knot theory, but most are written from the topological view point, and so the papers referred above are about the only ones with differential geometry results.

Chapter 2
Local Theory of Surfaces in the Space

2.1 Concept of Surfaces in the Space

Just as plane curves are given by

$$y = f(x), \tag{2.1.1}$$

$$F(x, y) = 0, \tag{2.1.2}$$

$$x = x(t), \quad y = y(t), \tag{2.1.3}$$

surfaces in the space are expressed as

$$z = f(x, y), \tag{2.1.4}$$

$$F(x, y, z) = 0, \tag{2.1.5}$$

$$x = x(u, v), \quad y = y(u, v), \quad z = z(u, v). \tag{2.1.6}$$

(2.1.4) is a special case of (2.1.5) and (2.1.6). In fact, setting $F(x, y, z) = f(x, y) - z$, (2.1.4) can be written in the form of (2.1.5), and by setting $x = u$, $y = v$, $z = f(u, v)$, it can be written in the form of (2.1.6). The graph of $z = f(x, y)$ is a smooth surface if f is differentiable, but in order for the graph of (2.1.5) or (2.1.6) to be a smooth surface, the following condition must be satisfied.

First, for $F(x, y, z) = 0$, we have learned in calculus that the tangent plane at a point (x_0, y_0, z_0) on this surface, is given by

$$\begin{aligned} F_x(x_0, y_0, z_0)(x - x_0) + F_y(x_0, y_0, z_0)(y - y_0) \\ + F_z(x_0, y_0, z_0)(z - z_0) = 0. \end{aligned} \tag{2.1.7}$$

For (2.1.7) to define a plane, at least one of $F_x(x_0, y_0, z_0)$, $F_y(x_0, y_0, z_0)$, $F_z(x_0, y_0, z_0)$ must be nonzero. For example, let

$$F_z(x_0, y_0, z_0) \neq 0. \tag{2.1.8}$$

The original version of the chapter was revised: Belated corrections have been incorporated. The correction to the chapter is available at https://doi.org/10.1007/978-981-15-1739-6_6

S. Kobayashi, *Differential Geometry of Curves and Surfaces*,
Springer Undergraduate Mathematics Series,
https://doi.org/10.1007/978-981-15-1739-6_2

Then by the implicit function theorem, z can be written as a function of x and y in a neighborhood of (x_0, y_0):

$$z = f(x, y). \tag{2.1.9}$$

The set of points defined by $F(x, y, z) = 0$ will be a surface if and only if it is not an empty set, and at each point (x_0, y_0, z_0) where $F(x_0, y_0, z_0) = 0$, the total differential $dF = F_x dx + F_y dy + F_z dz$ is nonzero, i.e., the vector (F_x, F_y, F_z) is nonzero.

For a surface defined by (2.1.6), we learned in calculus that the tangent plane at a point $(x(u_0, v_0), y(u_0, v_0), z(u_0, v_0))$ is given by

$$\begin{vmatrix} x - x_0 & y - y_0 & z - z_0 \\ x_u(u_0, v_0) & y_u(u_0, v_0) & z_u(u_0, v_0) \\ x_v(u_0, v_0) & y_v(u_0, v_0) & z_v(u_0, v_0) \end{vmatrix} = 0, \tag{2.1.10}$$

where

$$x_0 = x(u_0, v_0), \quad y_0 = y(u_0, v_0), \quad z_0 = z(u_0, v_0).$$

Expand (2.1.10) by the first row. For (2.1.10) to be a plane, at least one of the coefficients of $x - x_0$, $y - y_0$, and $z - z_0$, namely one of the three determinants

$$\begin{vmatrix} y_u(u_0, v_0) & z_u(u_0, v_0) \\ y_v(u_0, v_0) & z_v(u_0, v_0) \end{vmatrix}, \quad \begin{vmatrix} z_u(u_0, v_0) & x_u(u_0, v_0) \\ z_v(u_0, v_0) & x_v(u_0, v_0) \end{vmatrix}, \quad \begin{vmatrix} x_u(u_0, v_0) & y_u(u_0, v_0) \\ x_v(u_0, v_0) & y_v(u_0, v_0) \end{vmatrix} \tag{2.1.11}$$

must be nonzero. For example, let the third determinant $\neq 0$. By the implicit function theorem z will be written as a function of x and y

$$z = f(x, y) \tag{2.1.12}$$

in a neighborhood of (x_0, y_0). Thus (2.1.6) defines a surface if and only if the determinants in (2.1.11) do not vanish simultaneously.

Summarizing the above, we give a definition of surface in the space. The most convenient method is the parametric representation (2.1.6), as in the case of curves. In (2.1.6), $x(u, v)$, $y(u, v)$, and $z(u, v)$ are functions three times differentiable, defined in a region D of plane (u, v). When the Jacobian matrix

$$\begin{bmatrix} x_u & y_u & z_u \\ x_v & y_v & z_v \end{bmatrix} \tag{2.1.13}$$

has rank 2 everywhere on D (so, the determinant of one of the matrices made by two of the above columns is not 0), (2.1.6) defines a surface in the space. When (u, v) moves within the region D, we call the figure we get from $(x(u, v), y(u, v), z(u, v))$ a **piece of surface**. When a set M in the space is a union of (possibly infinitely many) surfaces patches, we call M a **surface**. Thus a piece of surface is a special surface and is commonly called a surface. When we are talking about a neighborhood of a point in a surface, we are just talking about one piece of surface. This chapter is mostly about local differential geometry restricted to one piece of surface. We end this section by introducing some examples of surfaces.

Example 2.1.1 (Sphere : $x^2 + y^2 + z^2 = a^2$)

A parametric representation is given by

$$x = u, \quad y = v, \quad z = \sqrt{a^2 - u^2 - v^2}. \tag{2.1.14}$$

When (u, v) moves within region $D = \{(u, v); u^2 + v^2 < a^2\}$, (x, y, z) moves on the upper hemisphere where $z > 0$. Using a number of these hemispheres, we are able to cover the whole sphere. However, when calculating, another parameteric representation

$$x = a \cos u \cos v, \quad y = a \cos u \sin v, \quad z = a \sin u \tag{2.1.15}$$

is more convenient (see Figure 2.1.1). When (u, v) moves within the region $\{(u, v); |u| < \frac{\pi}{2}\}$, (x, y, z) covers on the sphere except the north and south pole. However, the correspondence is not one-to-one. (Changing v by 2π does not change

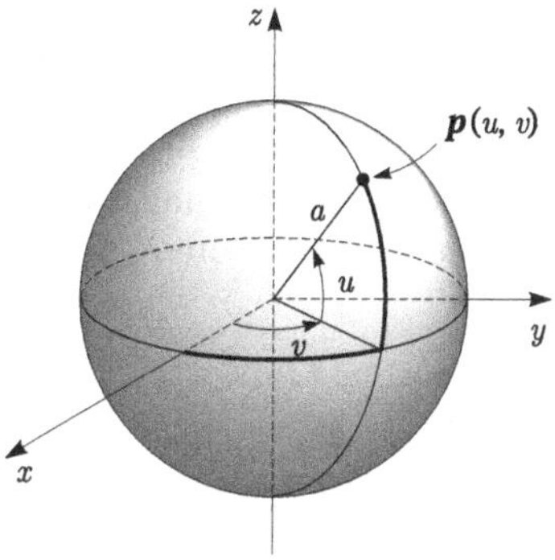

Fig. 2.1.1

(x, y, z).) In order to make it one-to-one, let the region be $D = \{(u, v); |u| < \frac{\pi}{2}, |v| < \pi\}$, and delete both poles and the international date line. (The real date line is not straight at 180°.)

In this way, if we delete the international date line and map the remainder to a rectangle D, it becomes just like a world map. Two world maps, one obtained by deleting a meridian through the Pacific and the other obtained by deleting a meridian through the Atlantic, cover the whole globe except the poles. We need two more maps with the north or south pole in the center to cover a sphere with pieces of surface. ♦

Example 2.1.2 (Ellipsoid : $\frac{x^2}{a^2} + \frac{y^2}{b^2} + \frac{z^2}{c^2} = 1$)

As in Example 2.1.1, we can cover the entire surface except the two points at the top and bottom if we let

$$x = a \cos u \cos v, \quad y = b \cos u \sin v, \quad z = c \sin u \tag{2.1.16}$$

(Figure 2.1.2). ♦

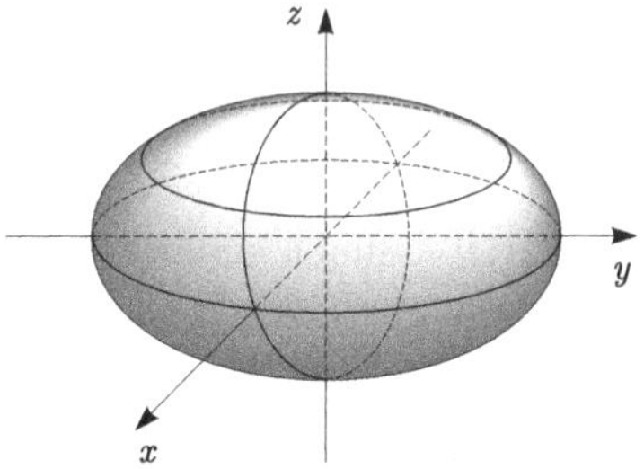

Fig. 2.1.2

Example 2.1.3 (Hyperboloid of one sheet : $\frac{x^2}{a^2} + \frac{y^2}{b^2} - \frac{z^2}{c^2} = 1$*)*
In this case, let

$$x = a \cosh u \cos v, \quad y = b \cosh u \sin v, \quad z = c \sinh u. \tag{2.1.17}$$

Then, when (u, v) moves on the plane without restrictions, (x, y, z) completely covers the hyperboloid of one sheet. However, by changing v to 2π, (x, y, z) does not change. So this correspondence is not one-to-one (Figure 2.1.3). ♦

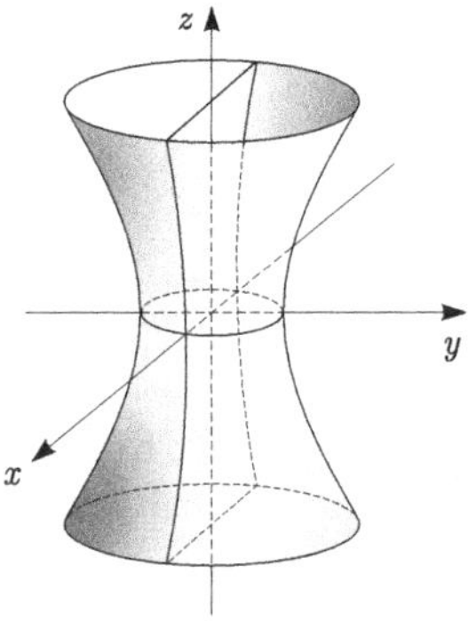

Fig. 2.1.3

Example 2.1.4 (Hyperboloid of two sheets : $\frac{x^2}{a^2} + \frac{y^2}{b^2} - \frac{z^2}{c^2} = -1$*)*
In this case, let

$$x = a \sinh u \cos v, \quad y = b \sinh u \sin v, \quad z = c \cosh u \tag{2.1.18}$$

and the region be $D = \{(u, v); u > 0\}$. Then we get a correspondence (not one-to-one) with the top half surface of the hyperboloid without the vertex. To get the bottom half, let $z = -c \cosh u$ (Figure 2.1.4). ♦

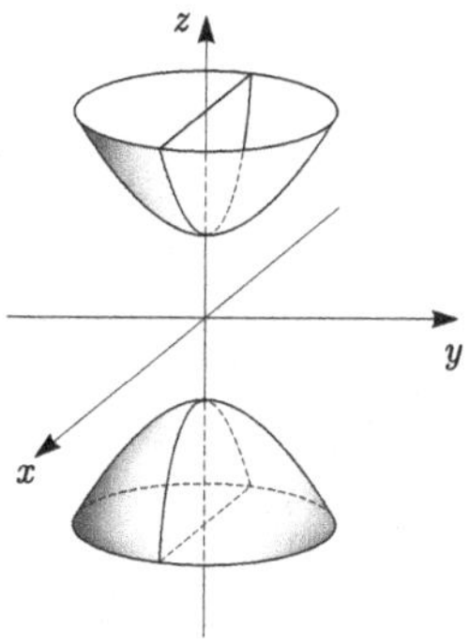

Fig. 2.1.4

Example 2.1.5 (Elliptic paraboloid : $z = \frac{x^2}{a^2} + \frac{y^2}{b^2}$)
In this case, of course,

$$x = u, \quad y = v, \quad z = \frac{u^2}{a^2} + \frac{v^2}{b^2} \tag{2.1.19}$$

is the simplest (Figure 2.1.5). ♦

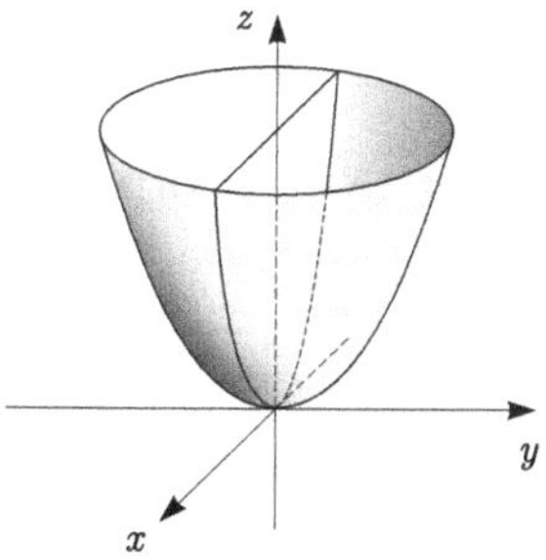

Fig. 2.1.5

Example 2.1.6 (Hyperbolic paraboloid : $z = -\frac{x^2}{a^2} + \frac{y^2}{b^2}$)
In this case, let

$$x = u, \quad y = v, \quad z = -\frac{u^2}{a^2} + \frac{v^2}{b^2} \tag{2.1.20}$$

(Figure 2.1.6). ♦

The examples above are all quadric surfaces. Next we will look at a doughnut shaped surface.

Example 2.1.7 (Torus)

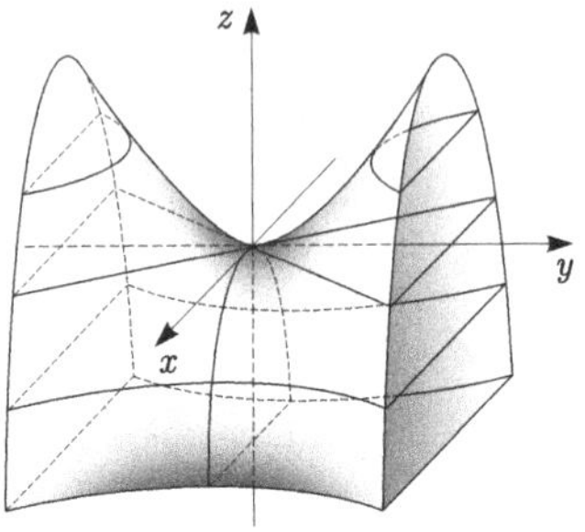

Fig. 2.1.6

This is a surface shaped like a doughnut or a life-saver. As in Figures 2.1.7 and 2.1.8, draw a circle on a x-axis so that its center is far enough from the origin, and revolve it around the z-axis to get this surface. Taking the center far enough from the origin means $r < R$ in Figure 2.1.7. The circle in Figure 2.1.7 is

$$x = R + r\cos u, \quad z = r\sin u, \tag{2.1.21}$$

so the equation of the torus in Figure 2.1.8 is

$$x = (R + r\cos u)\cos v, \quad y = (R + r\cos u)\sin v, \quad z = r\sin u. \tag{2.1.22}$$

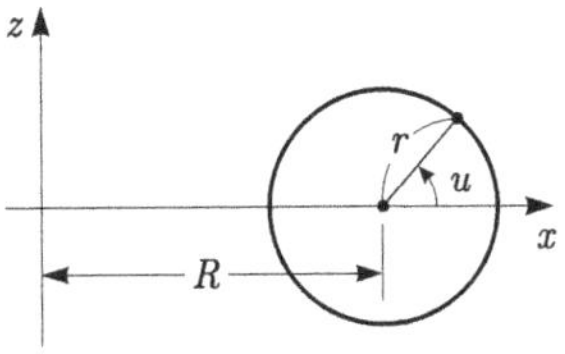

Fig. 2.1.7

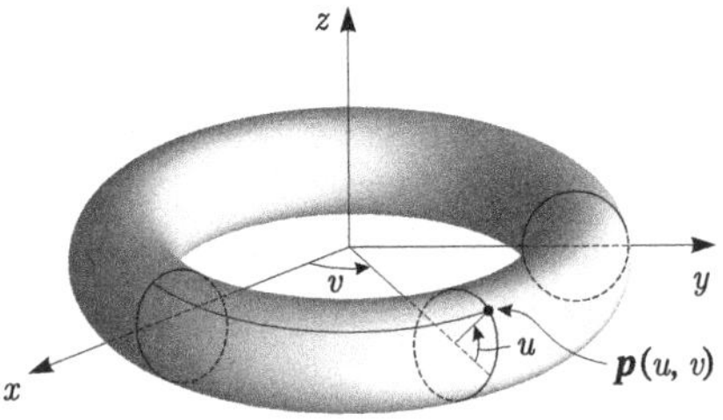

Fig. 2.1.8

In this case, D is the entire (u, v)-plane and covers the whole surface. But changing u to $u + 2\pi$ and v to $v + 2\pi$ does not change (x, y, z), so the correspondence is not one-to-one. ♦

Generalizing the above example we have

Example 2.1.8 (Surface of revolution)

On the (x, z)-plane, let

$$x = f(u), \quad z = g(u) \tag{2.1.23}$$

be a curve not intersecting the z-axis. Revolving it around the z-axis we get a surface with the equation

$$x = f(u)\cos v, \quad y = f(u)\sin v, \quad z = g(u). \tag{2.1.24}$$

♦

Problem 2.1.1 For each of the following surfaces, determine to which of the above examples it is equal.

(i) $x = au\cos v, \quad y = bu\sin v, \quad z = u^2.$

(ii) $x = au\cosh v, \quad y = bu\sinh v, \quad z = u^2.$

(iii) $x = a(u+v), \quad y = b(u-v), \quad z = 4uv.$

(iv) $x = a\dfrac{u-v}{u+v}, \quad y = b\dfrac{uv+1}{u+v}, \quad z = c\dfrac{uv-1}{u+v}.$

2.2 Fundamental Forms and Curvatures

As in the previous section, we will consider a parametric representation of a surface. As in the case of curves, we use vector notation and write a surface as follows,

$$\boldsymbol{p}(u,v) = (x(u,v), y(u,v), z(u,v)). \tag{2.2.1}$$

For the moment, we do not have to specify the region D where (u,v) moves. Now fixing v, say $v = b$, and changing u only, we get a curve $\boldsymbol{p}(u,b)$ on the surface. The

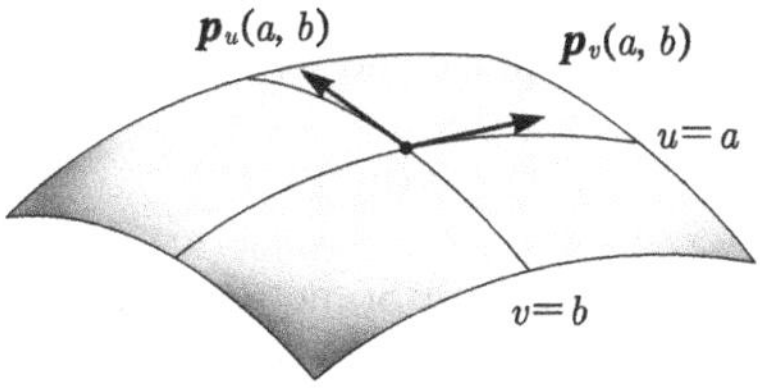

Fig. 2.2.1

velocity vector of this curve at point $\boldsymbol{p}(a,b)$ is given by

$$\boldsymbol{p}_u(a,b), \quad \text{where } \boldsymbol{p}_u = \frac{\partial \boldsymbol{p}}{\partial u}. \tag{2.2.2}$$

Also by fixing u at $u = a$ and changing v we get a curve $\boldsymbol{p}(a,v)$. Its velocity vector at point $\boldsymbol{p}(a,b)$ is given by

$$\boldsymbol{p}_v(a,b), \quad \text{where } \boldsymbol{p}_v = \frac{\partial \boldsymbol{p}}{\partial v}. \tag{2.2.3}$$

If we write out the components of the vectors $\boldsymbol{p}_u(a, b)$ and $\boldsymbol{p}_v(a, b)$, we have

$$\begin{aligned} \boldsymbol{p}_u(a, b) &= (x_u(a, b), y_u(a, b), z_u(a, b)), \\ \boldsymbol{p}_v(a, b) &= (x_v(a, b), y_v(a, b), z_v(a, b)), \end{aligned} \tag{2.2.4}$$

so for the matrix (2.1.13) to have rank 2 in the definition of a surface, vectors $\boldsymbol{p}_u(a, b)$ and $\boldsymbol{p}_v(a, b)$ have to be linearly independent. Thus, the tangent space of the surface $\boldsymbol{p}(u, v)$ at point $\boldsymbol{p}(a, b)$ will be spanned by $\boldsymbol{p}_u(a, b)$ and $\boldsymbol{p}_v(a, b)$. So far, we have been fixing (a, b) but by moving (a, b) we get 2 vector fields $\boldsymbol{p}_u$ and $\boldsymbol{p}_v$ on the surface (this means that at each point on the surface, $\boldsymbol{p}_u$ and $\boldsymbol{p}_v$ define tangent vectors). Let

$$E = \boldsymbol{p}_u \cdot \boldsymbol{p}_u, \quad F = \boldsymbol{p}_u \cdot \boldsymbol{p}_v = \boldsymbol{p}_v \cdot \boldsymbol{p}_u, \quad G = \boldsymbol{p}_v \cdot \boldsymbol{p}_v \tag{2.2.5}$$

be the inner products between $\boldsymbol{p}_u$ and $\boldsymbol{p}_v$, and consider the symmetric matrix

$$\begin{bmatrix} E & F \\ F & G \end{bmatrix} = \begin{bmatrix} \boldsymbol{p}_u \cdot \boldsymbol{p}_u & \boldsymbol{p}_u \cdot \boldsymbol{p}_v \\ \boldsymbol{p}_v \cdot \boldsymbol{p}_u & \boldsymbol{p}_v \cdot \boldsymbol{p}_v \end{bmatrix}. \tag{2.2.6}$$

Since $\boldsymbol{p}_u$ and $\boldsymbol{p}_v$ form a basis of the tangent plane at each point on the surface, each tangent vector can be written as a linear combination of $\boldsymbol{p}_u$ and $\boldsymbol{p}_v$,

$$\xi \boldsymbol{p}_u + \eta \boldsymbol{p}_v. \tag{2.2.7}$$

The square of the length of this vector is given by

$$(\xi \boldsymbol{p}_u + \eta \boldsymbol{p}_v) \cdot (\xi \boldsymbol{p}_u + \eta \boldsymbol{p}_v) = E\xi^2 + 2F\xi\eta + G\eta^2. \tag{2.2.8}$$

Let

$$u = u(t), \quad v = v(t) \tag{2.2.9}$$

be a curve in a region on the (u, v)-plane and let

$$\boldsymbol{p}(t) = \boldsymbol{p}(u(t), v(t)) \tag{2.2.10}$$

be the corresponding curve on the surface. Sometimes we identify (2.2.9) and (2.2.10) and call it a curve $(u(t), v(t))$ on the surface $\boldsymbol{p}(u, v)$. Now, the tangent vector at each point on this curve $\boldsymbol{p}(t)$ is given by

$$\frac{d\boldsymbol{p}}{dt} = \frac{\partial \boldsymbol{p}}{\partial u}\frac{du}{dt} + \frac{\partial \boldsymbol{p}}{\partial v}\frac{dv}{dt} = \boldsymbol{p}_u \frac{du}{dt} + \boldsymbol{p}_v \frac{dv}{dt}. \tag{2.2.11}$$

So the square of its length is equal to

$$\frac{d\boldsymbol{p}}{dt} \cdot \frac{d\boldsymbol{p}}{dt} = E\left(\frac{du}{dt}\right)^2 + 2F\frac{du}{dt}\frac{dv}{dt} + G\left(\frac{dv}{dt}\right)^2. \tag{2.2.12}$$

When t moves from α to β the length of this curve is given by

$$\int_{\alpha}^{\beta} \sqrt{E\left(\frac{du}{dt}\right)^2 + 2F\frac{du}{dt}\frac{dv}{dt} + G\left(\frac{dv}{dt}\right)^2}\, dt. \tag{2.2.13}$$

In view of (2.2.12) and (2.2.13) we are led to consider the expression

$$\mathbf{I} = E\, dudu + 2F\, dudv + G\, dvdv \tag{2.2.14}$$

and call it the **first fundamental form** of the surface $\boldsymbol{p}(u, v)$.

In order to understand (2.2.14) better we have to read more advanced books on differential geometry. For now, understanding it in the following way is enough. As in calculus, when t changes by a very small Δt to $t + \Delta t$, let $u(t)$ change to $u(t) + \Delta u$ and $v(t)$ to $v(t) + \Delta v$. Then

$$\boldsymbol{p}(t + \Delta t) - \boldsymbol{p}(t) \sim \boldsymbol{p}_u(u(t), v(t))\Delta u + \boldsymbol{p}_v(u(t), v(t))\Delta v, \tag{2.2.15}$$

(where $\sim$ means "approximately equal" when Δu and Δv are small). Thus as t moves to $t + \Delta t$, $\boldsymbol{p}(t)$ moves to $\boldsymbol{p}(t + \Delta t)$, and the square of the distance between $\boldsymbol{p}(t)$ and $\boldsymbol{p}(t + \Delta t)$ is obtained by taking the inner product of (2.2.15) with itself, which is about

$$E(u(t), v(t))\Delta u\Delta u + 2F(u(t), v(t))\Delta u\Delta v + G(u(t), v(t))\Delta v\Delta v. \tag{2.2.16}$$

Thus (2.2.14) can be seen as the limit of (2.2.16) when Δt approaches 0.

When using (2.2.14) we will always change it into the form (2.2.12) or (2.2.13) so we do not have to worry about the meaning of (2.2.14). Since we are studying differential geometry as a continuation of calculus without using advanced mathematics, ambiguous points such as this may arise. However, readers should go on disregarding those points. In calculus also, we used symbols such as $f(x)dx$ but in actual calculations, we used it in the form of $f(x)\frac{dx}{dt}$ or $\int f(x)dx$. If we use only (2.2.12) or (2.2.13) in actual calculations, why do we have the expression like (2.2.14) ? Such expressions as (2.2.14) are convenient at times, for example, to memorize formulas and to foresee results in calculation. The history of mathematics shows that concepts, symbols, and equations which may be ambiguous at first but which are useful and lead to correct results get eventually justified. Formally, (2.2.15) can be written as

$$d\boldsymbol{p} = \boldsymbol{p}_u\, du + \boldsymbol{p}_v\, dv \tag{2.2.17}$$

so the first fundamental form can be written as

$$\mathbf{I} = d\boldsymbol{p} \cdot d\boldsymbol{p}. \tag{2.2.18}$$

Since (2.2.8) is the square of the length of (2.2.7), it is always positive unless $\xi = \eta = 0$. Therefore this means that, as a quadratic form, (2.2.14) is **a positive definite form**.

Since the vector product $\boldsymbol{p}_u \times \boldsymbol{p}_v$ is orthogonal to the two vectors $\boldsymbol{p}_u$ and $\boldsymbol{p}_v$, it is perpendicular to the tangent plane. Thus it is a normal vector. If we set

$$\boldsymbol{e} = \frac{\boldsymbol{p}_u \times \boldsymbol{p}_v}{|\boldsymbol{p}_u \times \boldsymbol{p}_v|}, \tag{2.2.19}$$

$\boldsymbol{e}$ is a normal vector with length 1. We define the **second fundamental form** of the surface by

$$\mathrm{II} = -d\boldsymbol{p} \cdot d\boldsymbol{e}. \tag{2.2.20}$$

This is the simplest definition, but it is not a good clear definition. Its geometric meaning will become clear later.

Since $\boldsymbol{e}$ is orthogonal to $\boldsymbol{p}_u$ and $\boldsymbol{p}_v$, we have

$$\boldsymbol{p}_u \cdot \boldsymbol{e} = 0, \quad \boldsymbol{p}_v \cdot \boldsymbol{e} = 0. \tag{2.2.21}$$

Partially differentiating this equation by u and v we have

$$\begin{aligned} \boldsymbol{p}_{uu} \cdot \boldsymbol{e} + \boldsymbol{p}_u \cdot \boldsymbol{e}_u = 0, &\quad \boldsymbol{p}_{uv} \cdot \boldsymbol{e} + \boldsymbol{p}_u \cdot \boldsymbol{e}_v = 0, \\ \boldsymbol{p}_{vu} \cdot \boldsymbol{e} + \boldsymbol{p}_v \cdot \boldsymbol{e}_u = 0, &\quad \boldsymbol{p}_{vv} \cdot \boldsymbol{e} + \boldsymbol{p}_v \cdot \boldsymbol{e}_v = 0. \end{aligned} \tag{2.2.22}$$

Thus we can define functions L, M and N by

$$\begin{aligned} L &= \boldsymbol{p}_{uu} \cdot \boldsymbol{e} = -\boldsymbol{p}_u \cdot \boldsymbol{e}_u, \\ M &= \boldsymbol{p}_{uv} \cdot \boldsymbol{e} = -\boldsymbol{p}_u \cdot \boldsymbol{e}_v, \\ M &= \boldsymbol{p}_{vu} \cdot \boldsymbol{e} = -\boldsymbol{p}_v \cdot \boldsymbol{e}_u, \\ N &= \boldsymbol{p}_{vv} \cdot \boldsymbol{e} = -\boldsymbol{p}_v \cdot \boldsymbol{e}_v. \end{aligned} \tag{2.2.23}$$

Using (2.2.23) we have

$$\begin{aligned} \mathrm{II} &= -d\boldsymbol{p} \cdot d\boldsymbol{e} \\ &= -(\boldsymbol{p}_u du + \boldsymbol{p}_v dv) \cdot (\boldsymbol{e}_u du + \boldsymbol{e}_v dv) \\ &= L\, dudu + 2M\, dudv + N\, dvdv. \end{aligned} \tag{2.2.24}$$

At each point on the surface, $\boldsymbol{p}_u$, $\boldsymbol{p}_v$ and $\boldsymbol{e}$ are linearly independent, so any vectors in the space can be represented as a linear combination of these three vectors. In particular, the vectors $\boldsymbol{p}_{uu}$, $\boldsymbol{p}_{uv}$, $\boldsymbol{p}_{vu}$, $\boldsymbol{p}_{vv}$, $\boldsymbol{e}_u$, and $\boldsymbol{e}_v$ can be written as linear combinations of $\boldsymbol{p}_u$, $\boldsymbol{p}_v$ and $\boldsymbol{e}$. Namely, we have

$$\begin{aligned} \boldsymbol{p}_{uu} &= \Gamma^u_{uu}\, \boldsymbol{p}_u + \Gamma^v_{uu}\, \boldsymbol{p}_v + L\, \boldsymbol{e}, \\ \boldsymbol{p}_{uv} &= \Gamma^u_{uv}\, \boldsymbol{p}_u + \Gamma^v_{uv}\, \boldsymbol{p}_v + M\, \boldsymbol{e}, \\ \boldsymbol{p}_{vu} &= \Gamma^u_{vu}\, \boldsymbol{p}_u + \Gamma^v_{vu}\, \boldsymbol{p}_v + M\, \boldsymbol{e}, \\ \boldsymbol{p}_{vv} &= \Gamma^u_{vv}\, \boldsymbol{p}_u + \Gamma^v_{vv}\, \boldsymbol{p}_v + N\, \boldsymbol{e}, \\ \boldsymbol{e}_u &= \frac{FM - GL}{EG - F^2}\, \boldsymbol{p}_u + \frac{FL - EM}{EG - F^2}\, \boldsymbol{p}_v, \\ \boldsymbol{e}_v &= \frac{FN - GM}{EG - F^2}\, \boldsymbol{p}_u + \frac{FM - EN}{EG - F^2}\, \boldsymbol{p}_v. \end{aligned} \tag{2.2.25}$$

Here, $\Gamma^u_{uu}, \Gamma^v_{uu}, \ldots, \Gamma^v_{vv}$ are defined by the above equation, and what we need to prove is that the other coefficients are represented by E, F, G, L, M, and N. $\Gamma^u_{uu}, \ldots, \Gamma^v_{vv}$ defined above are called **Christoffel's symbols**. The first four lines of (2.2.25) are called the **Gauss' equations** and the last two line are called the **Weingarten's equations**.

In the Gauss' equations, we can see from (2.2.23) that the coefficients of $\boldsymbol{e}$ are L, M, M, and N. To prove the Weingarten's equations, first differentiate $\boldsymbol{e} \cdot \boldsymbol{e} = 1$ by u to get $\boldsymbol{e}_u \cdot \boldsymbol{e} = 0$. From this we know that $\boldsymbol{e}_u$ is orthogonal to $\boldsymbol{e}$. Thus we should be able to write it as

$$\boldsymbol{e}_u = A\,\boldsymbol{p}_u + B\,\boldsymbol{p}_v,$$

and from the inner product of the above equation with $\boldsymbol{p}_u$ and $\boldsymbol{p}_v$, using (2.2.23) and (2.2.5) we have

$$-L = \boldsymbol{p}_u \cdot \boldsymbol{e}_u = EA + FB,$$
$$-M = \boldsymbol{p}_v \cdot \boldsymbol{e}_u = FA + GB.$$

Solving these for A and B we have

$$A = \frac{FM - GL}{EG - F^2}, \quad B = \frac{FL - EM}{EG - F^2}.$$

The proof for $\boldsymbol{e}_v$ is similar. It is known that $\Gamma^u_{uu}, \cdots$ are represented by E, F, G and its partial derivative, but since the equations are complicated we omit them here. We may say that (2.2.25) is an analogue of Frenet-Serret's formula for curves. Later, we will prove the equations in a better form.

In order to understand the geometric meaning of the second fundamental form, we fix a unit vector $\boldsymbol{a}$ and define a function f on the surface by

$$f(u, v) = \boldsymbol{a} \cdot \boldsymbol{p}(u, v). \tag{2.2.26}$$

From Figure 2.2.2, we can see that when we rotate the whole space so that vector $\boldsymbol{a}$ points upwards, $f(u, v)$ becomes the height of $\boldsymbol{p}(u, v)$. Thus f is a function representing the height in the direction $\boldsymbol{a}$.

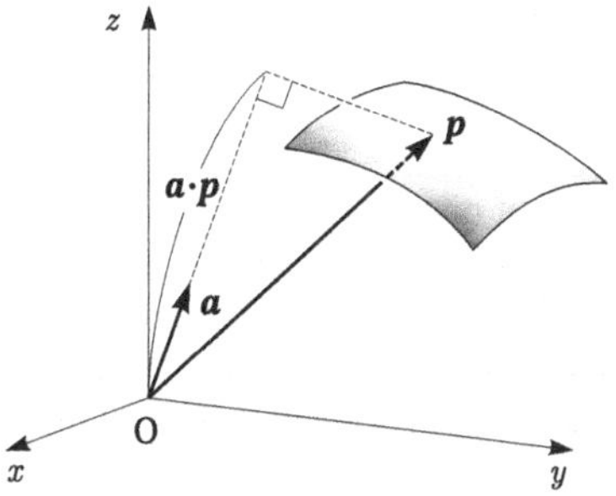

Fig. 2.2.2

Now, fix a point $\boldsymbol{p}_0 = \boldsymbol{p}(u_0, v_0)$ on a surface, and let $\boldsymbol{a}$ be the normal vector $\boldsymbol{a} = \boldsymbol{e}(u_0, v_0)$ at that point. Since $\boldsymbol{a} \cdot \boldsymbol{p}_u(u_0, v_0) = \boldsymbol{a} \cdot \boldsymbol{p}_v(u_0, v_0) = 0$,

$$df = d(\boldsymbol{a} \cdot \boldsymbol{p}) = \boldsymbol{a} \cdot \boldsymbol{p}_u du + \boldsymbol{a} \cdot \boldsymbol{p}_v dv \tag{2.2.27}$$

becomes 0 at $p_0 = p(u_0, v_0)$. (Geometrically speaking, if we assume a is a vector in the vertical direction, the tangent plane at p_0 is orthogonal to a so it is horizontal and this means that the height of the surface has reached the **critical value** at p_0.) Since df is 0 at p_0, from checking the **Hessian** of f we can see the behavior of f in a neighborhood of p_0. The Hessian we obtain by partially differentiating f twice is a symmetric matrix

$$H_f = \begin{bmatrix} f_{uu} & f_{uv} \\ f_{vu} & f_{vv} \end{bmatrix}. \tag{2.2.28}$$

Using (2.2.23), (2.2.26), and $a = e(u_0, v_0)$ we get

$$H_f(u_0, v_0) = \begin{bmatrix} L(u_0, v_0) & M(u_0, v_0) \\ M(u_0, v_0) & N(u_0, v_0) \end{bmatrix}. \tag{2.2.29}$$

Thus if we assume that the second fundamental form

$$\mathrm{II} = L\,du\,du + 2M\,du\,dv + N\,dv\,dv \tag{2.2.30}$$

is positive-definite at $p(u_0, v_0)$, so is the Hessian H_f at (u_0, v_0). Hence as we learned in calculus, the function f is concave upwards at (u_0, v_0). This shows that the surface is concave in the direction of the normal vector a at $p(u_0, v_0)$ (convex in the opposite direction of a). Also, if II of (2.2.30) is negative-definite at $p(u_0, v_0)$, the surface is convex in the direction of a at $p(u_0, v_0)$ (see Figure 2.2.3).

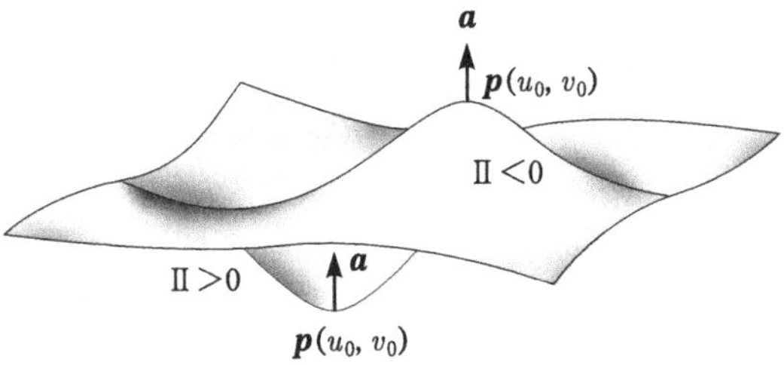

Fig. 2.2.3

When II is indefinite, i.e., when $LN - M^2 < 0$, we get Figure 2.2.4. For $LN - M^2 = 0$, as in the case of a function of one variable, we cannot say anything in general.

The positive and negative definite cases of the second fundamental form II are essentially the same, because when we change the normal vector from e to $-e$, the sign of II also changes. We can see this in Figure 2.2.3; the part of the surface where $\mathrm{II} > 0$ is of the same shape as the part where $\mathrm{II} < 0$. Summarizing, we have the following theorem.

Theorem 2.2.1 *At a point where the second fundamental form*

$$\mathrm{II} = L\,du\,du + 2M\,du\,dv + N\,dv\,dv$$

is definite (either positive or negative), i.e., a point where $LN - M^2 > 0$, *the surface is convex (concave if it is looked at from the opposite direction), and at a point where* II *is indefinite, i.e.,* $LN - M^2 < 0$, *it is saddle shaped.*

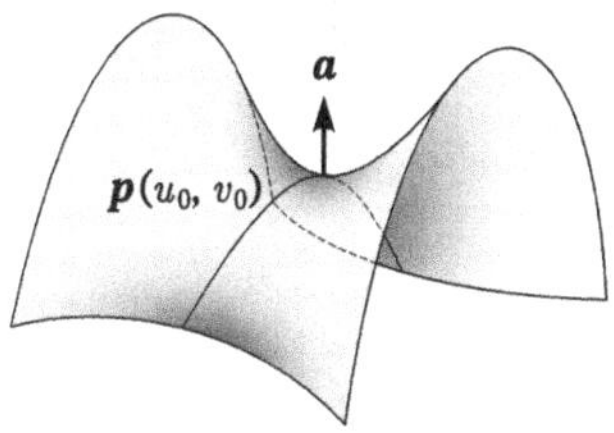

Fig. 2.2.4

In order to see the quantitative relationship between the second fundamental form **II** and the way the surface is curved, we calculate the curvature of a curve on the surface. Consider a curve given by $(u(s), v(s))$ as a space curve $\boldsymbol{p}(s) = \boldsymbol{p}(u(s), v(s))$ and calculate its curvature κ. For convenience let s be a parameter such that $\left|\frac{d\boldsymbol{p}}{ds}\right| = 1$. In (1.4.3), we defined the length of the vector $\boldsymbol{p}''(s)$ as the curvature of the curve $\boldsymbol{p}(s)$. Since $\boldsymbol{p}'(s)$ is a tangent vector of the curve $\boldsymbol{p}(s)$, it is also tangent to the surface $\boldsymbol{p}(u, v)$. However, generally $\boldsymbol{p}''(s)$ is not tangent to the surface $\boldsymbol{p}(u, v)$, so by decomposing it into its tangential component $\boldsymbol{k}_g$ and its normal component $\boldsymbol{k}_n$, we write

$$\boldsymbol{p}''(s) = \boldsymbol{k}_g + \boldsymbol{k}_n. \tag{2.2.31}$$

We call $\boldsymbol{k}_g$ the **geodesic curvature vector** of the curve $\boldsymbol{p}(s)$ on the surface, and $\boldsymbol{k}_n$ the **normal curvature vector**. We will look into $\boldsymbol{k}_g$ in more detail later. Here we consider $\boldsymbol{k}_n$. Since $\boldsymbol{k}_n$ is a normal vector, using the unit normal vector $\boldsymbol{e}$ we can write it as

$$\boldsymbol{k}_n = \kappa_n \, \boldsymbol{e}. \tag{2.2.32}$$

We call κ_n the **normal curvature** of the curve $\boldsymbol{p}(s)$ on the surface $\boldsymbol{p}(u, v)$. From (2.2.32), (2.2.31), and (2.2.22) (or from (2.2.24) we get,

$$\begin{aligned}
\kappa_n &= \kappa_n \, \boldsymbol{e} \cdot \boldsymbol{e} = \boldsymbol{k}_n \cdot \boldsymbol{e} = (\boldsymbol{p}'' - \boldsymbol{k}_g) \cdot \boldsymbol{e} = \boldsymbol{p}'' \cdot \boldsymbol{e} = -\boldsymbol{p}' \cdot \boldsymbol{e}' \\
&= -\left(\boldsymbol{p}_u \frac{du}{ds} + \boldsymbol{p}_v \frac{dv}{ds}\right) \cdot \left(\boldsymbol{e}_u \frac{du}{ds} + \boldsymbol{e}_v \frac{dv}{ds}\right) \\
&= L \frac{du}{ds}\frac{du}{ds} + 2M \frac{du}{ds}\frac{dv}{ds} + N \frac{dv}{ds}\frac{dv}{ds}.
\end{aligned} \tag{2.2.33}$$

According to the above equation, we know that $\kappa_n(s)$ is determined by the second fundamental form **II** of the surface at $\boldsymbol{p}(u(s), v(s))$ and the vector $\boldsymbol{p}'(s)$ only.

Thus, for a unit tangent vector

$$\boldsymbol{w} = \xi \, \boldsymbol{p}_u(u_0, v_0) + \eta \, \boldsymbol{p}_v(u_0, v_0) \tag{2.2.34}$$

at a point $\boldsymbol{p}_0 = \boldsymbol{p}(u_0, v_0)$ of the surface, we consider

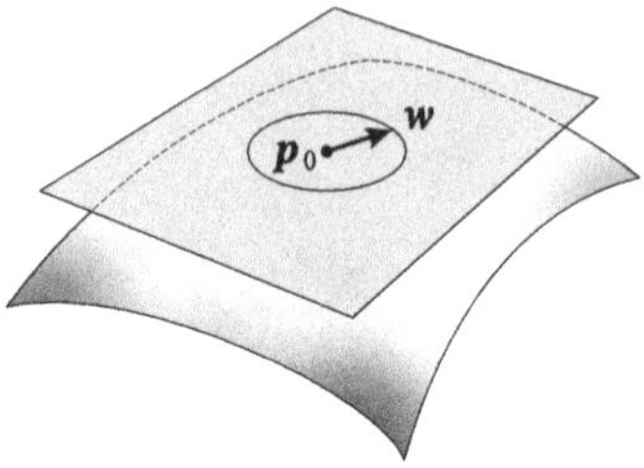

Fig. 2.2.5

$$\mathbf{I\!I}(w, w) = L\,\xi^2 + 2M\,\xi\,\eta + N\,\eta^2. \tag{2.2.35}$$

With the notation (2.2.35), the normal curvature (2.2.33) can be written as

$$\kappa_n(s) = \mathbf{I\!I}(p'(s), p'(s)). \tag{2.2.36}$$

Let us calculate the maximum and minimum value of $\mathbf{I\!I}(w, w)$ when w moves on the unit circle in the tangent plane at p_0. This is the same as calculating the maximum and minimum of (2.2.35) under the constraint

$$|w|^2 = E\,\xi^2 + 2F\,\xi\,\eta + G\,\eta^2 = 1. \tag{2.2.37}$$

Instead, we may calculate the maximum and minimum of the function

$$\lambda = \frac{L\,\xi^2 + 2M\,\xi\,\eta + N\,\eta^2}{E\,\xi^2 + 2F\,\xi\,\eta + G\,\eta^2} \tag{2.2.38}$$

under the condition $(\xi, \eta) \neq (0, 0)$. This is because when w is multiplied by c, both the numerator and the denominator of λ is multiplied by c^2 and λ remains invariant. Consider

$$L\,\xi^2 + 2M\,\xi\,\eta + N\,\eta^2 - \lambda(E\,\xi^2 + 2F\,\xi\,\eta + G\,\eta^2) = 0. \tag{2.2.39}$$

Differentiate by ξ and η. Then by setting $\frac{\partial \lambda}{\partial \xi} = \frac{\partial \lambda}{\partial \eta} = 0$ we get

$$\begin{aligned}(L - \lambda E)\,\xi + (M - \lambda F)\,\eta &= 0,\\ (M - \lambda F)\,\xi + (N - \lambda G)\,\eta &= 0.\end{aligned} \tag{2.2.40}$$

Instead of looking for a solution $(\xi, \eta) \neq (0, 0)$ of (2.2.40) and calculating λ, find a λ for which (2.2.40) has a solution $(\xi, \eta) \neq (0, 0)$. This amounts to solving the determinant equation

$$\begin{vmatrix} L - \lambda E & M - \lambda F \\ M - \lambda F & N - \lambda G \end{vmatrix} = 0. \tag{2.2.41}$$

Expand (2.2.41) into a polynomial of degree 2 in λ:

$$(EG - F^2)\lambda^2 - (EN + GL - 2FM)\lambda + LN - M^2 = 0 \tag{2.2.42}$$

and let its solutions be $\lambda = \kappa_1, \kappa_2$. Then from the relationship between the roots and the coefficients of the polynomial (2.2.42), we get

$$\kappa_1\kappa_2 = \frac{LN - M^2}{EG - F^2}, \quad \frac{1}{2}(\kappa_1 + \kappa_2) = \frac{EN + GL - 2FM}{2(EG - F^2)}. \tag{2.2.43}$$

Let

$$K = \kappa_1\kappa_2, \qquad H = \frac{1}{2}(\kappa_1 + \kappa_2), \tag{2.2.44}$$

and call K the **Gaussian curvature** and H the **mean curvature**. When $K \equiv 0$, the surface is said to be **flat**, and when $H \equiv 0$ the surface is a **minimal surface**. We call κ_1 and κ_2 the **principal curvatures** and call the direction of unit vectors $\boldsymbol{w}_1$ and $\boldsymbol{w}_2$ for which $\mathrm{II}(\boldsymbol{w}_1, \boldsymbol{w}_1) = \kappa_1$ and $\mathrm{II}(\boldsymbol{w}_2, \boldsymbol{w}_2) = \kappa_2$ the **principal directions** at point $\boldsymbol{p}_0$.

In order to study properties of the principal directions, let $\lambda = \kappa_i$ in (2.2.40) and let a vector $\boldsymbol{w}_i$ in the corresponding principal direction have components (ξ_i, η_i). Then we have

$$\begin{aligned} L\,\xi_i + M\,\eta_i &= \kappa_i(E\,\xi_i + F\,\eta_i), \\ M\,\xi_i + N\,\eta_i &= \kappa_i(F\,\xi_i + G\,\eta_i) \end{aligned} \tag{2.2.45}$$

for $i = 1, 2$. Rewrite (2.2.45) in matrix notation as

$$\begin{bmatrix} L & M \\ M & N \end{bmatrix} \begin{bmatrix} \xi_i \\ \eta_i \end{bmatrix} = \kappa_i \begin{bmatrix} E & F \\ F & G \end{bmatrix} \begin{bmatrix} \xi_i \\ \eta_i \end{bmatrix} \qquad (i = 1, 2), \tag{2.2.46}$$

and get

$$\begin{aligned} &\kappa_2(\xi_1, \eta_1) \begin{bmatrix} E & F \\ F & G \end{bmatrix} \begin{bmatrix} \xi_2 \\ \eta_2 \end{bmatrix} \\ &= (\xi_1, \eta_1) \begin{bmatrix} L & M \\ M & N \end{bmatrix} \begin{bmatrix} \xi_2 \\ \eta_2 \end{bmatrix} = (\xi_2, \eta_2) \begin{bmatrix} L & M \\ M & N \end{bmatrix} \begin{bmatrix} \xi_1 \\ \eta_1 \end{bmatrix} \\ &= \kappa_1(\xi_2, \eta_2) \begin{bmatrix} E & F \\ F & G \end{bmatrix} \begin{bmatrix} \xi_1 \\ \eta_1 \end{bmatrix} = \kappa_1(\xi_1, \eta_1) \begin{bmatrix} E & F \\ F & G \end{bmatrix} \begin{bmatrix} \xi_2 \\ \eta_2 \end{bmatrix}. \end{aligned} \tag{2.2.47}$$

(In the above calculation, the first and third equalities are from (2.2.46), the second and fourth equalities are from the symmetricity of the second fundamental form matrix and the first fundamental form matrix.) Thus, when $\kappa_1 \neq \kappa_2$,

$$(\xi_1, \eta_1) \begin{bmatrix} E & F \\ F & G \end{bmatrix} \begin{bmatrix} \xi_2 \\ \eta_2 \end{bmatrix} = 0, \tag{2.2.48}$$

(i.e., $E\xi_1\xi_2 + F(\xi_1\eta_2 + \eta_1\xi_2) + G\eta_1\eta_2 = 0$). This shows that when $\kappa_1 \neq \kappa_2$, the two principal directions are orthogonal to each other.

Next, let $\kappa_1 = \kappa_2$. Then under the condition (2.2.37), the maximum and the minimum of (2.2.35) are equal. So as $\boldsymbol{w}$ moves on the unit circle in the tangent plane at $\boldsymbol{p}_0$, $\mathrm{II}(\boldsymbol{w}, \boldsymbol{w})$ remains constant $\kappa = \kappa_1 = \kappa_2$. Thus (2.2.38) is a constant κ for for all $(\xi, \eta) \neq (0, 0)$ at $\boldsymbol{p}_0$. Clearing the denominator, we have

$$L\,\xi^2 + 2M\,\xi\,\eta + N\,\eta^2 = \kappa(E\,\xi^2 + 2F\,\xi\,\eta + G\,\eta^2) \tag{2.2.49}$$

for all (ξ, η) at $\boldsymbol{p}_0$. We call a point $\boldsymbol{p}_0$ where $\kappa_1 = \kappa_2$ an **umbilic point** of the surface. At an umbilic point, all directions become principal directions, so there are no two special directions. Geometrically, at that point the surface curves in the same way for all directions.

On the other hand, we call a point where $K > 0$ an **elliptic point**, where $K < 0$ a **hyperbolic point**, and where $K = 0$ a **parabolic point**.

The first fundamental form is always positive definite, so $EG - F^2 > 0$ and by equation (2.2.43), K has the same sign as $LN - M^2$. Thus, we can rewrite Theorem 2.2.1 as follows.

Theorem 2.2.2 *At a point where the Gaussian curvature K is positive, i.e., at an elliptic point, the surface is convex, and at a point where K is negative, i.e., at a hyperbolic point, it is saddle shaped.*

For a curve $\boldsymbol{p}(s) = \boldsymbol{p}(u(s), v(s))$ on a surface $\boldsymbol{p}(u, v)$, we defined the geodesic curvature vector $\boldsymbol{k}_g$ as the tangential component of $\boldsymbol{p}''(s)$ (see (2.2.31)). When $\boldsymbol{k}_g \equiv \boldsymbol{0}$ we call $\boldsymbol{p}(s)$ a **geodesic**. Thus when the acceleration vector $\boldsymbol{p}''(s)$ of $\boldsymbol{p}(s)$ is everywhere perpendicular to the surface, we call $\boldsymbol{p}(s)$ the geodesic. We will discuss geodesics in more detail later.

As in the case of a plane curve, using the unit normal vector $\boldsymbol{e}$ we define Gauss' spherical map, and consider its relationship with the Gaussian curvature. Consider the unit normal vector $\boldsymbol{e}$ defined in (2.2.19) as a point on the unit sphere with center at the origin of the space. Then for each point $\boldsymbol{p}(a, b)$ on the surface, there corresponds a point $\boldsymbol{e}(a, b)$ on the unit sphere. We call this correspondence **Gauss' spherical map**. For plane curves, the curvature at a point $\boldsymbol{p}_0$ on the curve was the ratio between the distance a point travels along the curve near $\boldsymbol{p}_0$ and the distance traveled on the unit circle by the corresponding point under the Gauss map (see (1.2.27)). We shall now explain that for a surface, the Gaussian curvature is the ratio between areas.

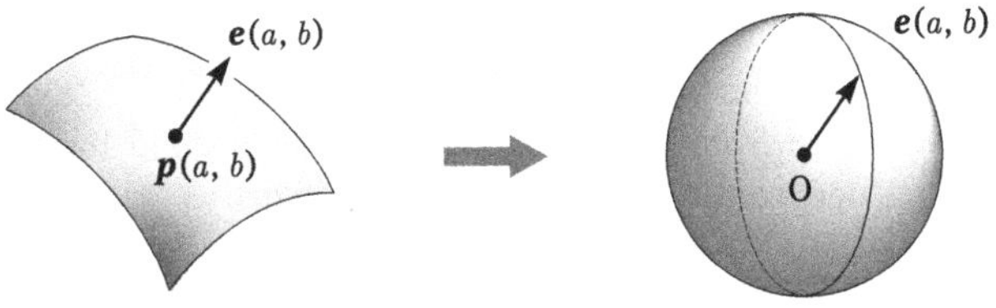

Fig. 2.2.6

First, consider the parallelogram spanned by vectors $\boldsymbol{p}_u(a, b)$ and $\boldsymbol{p}_v(a, b)$ on the tangent plane at point $\boldsymbol{p}(a, b)$. The area of this parallelogram is given by the length of the vector product

$$|\boldsymbol{p}_u(a, b) \times \boldsymbol{p}_v(a, b)| \tag{2.2.50}$$

of these two vectors. Similarly, for a point $\boldsymbol{e}(a, b)$ on a unit sphere, consider the parallelogram spanned by the two corresponding tangent vectors $\boldsymbol{e}_u(a, b)$ and $\boldsymbol{e}_v(a, b)$. Then its area is given by

$$|\boldsymbol{e}_u(a, b) \times \boldsymbol{e}_v(a, b)|. \tag{2.2.51}$$

Using the last two equations in (2.2.25), we have

$$\begin{aligned}
&\boldsymbol{e}_u \times \boldsymbol{e}_v \\
&= \left(\frac{FM-GL}{EG-F^2}\boldsymbol{p}_u + \frac{FL-EM}{EG-F^2}\boldsymbol{p}_v\right) \times \left(\frac{FN-GM}{EG-F^2}\boldsymbol{p}_u + \frac{FM-EN}{EG-F^2}\boldsymbol{p}_v\right) \\
&= \frac{(FM-GL)(FM-EN)-(FN-GM)(FL-EM)}{(EG-F^2)^2}\boldsymbol{p}_u \times \boldsymbol{p}_v \\
&= \frac{LN-M^2}{EG-F^2}\boldsymbol{p}_u \times \boldsymbol{p}_v.
\end{aligned} \tag{2.2.52}$$

Thus from (2.2.43) we have

$$\boldsymbol{e}_u \times \boldsymbol{e}_v = K(\boldsymbol{p}_u \times \boldsymbol{p}_v). \tag{2.2.53}$$

From this we get

$$|\boldsymbol{e}_u \times \boldsymbol{e}_v| = |K| \cdot |\boldsymbol{p}_u \times \boldsymbol{p}_v|, \tag{2.2.54}$$

and this shows that the ratio of the area of the two parallelograms are $|K|$. However, in (2.2.54), we cannot tell whether K is positive or negative. So getting back to (2.2.53), from the definition of $\boldsymbol{e}$ in (2.2.19) we see that $\boldsymbol{p}_u \times \boldsymbol{p}_v$ has the same direction as $\boldsymbol{e}$. Thus if K is positive, $\boldsymbol{e}_u \times \boldsymbol{e}_v$ is also a vector in the same direction as $\boldsymbol{e}$. This indicates that $(\boldsymbol{e}_u, \boldsymbol{e}_v, \boldsymbol{e})$ is the right-handed system like $(\boldsymbol{p}_u, \boldsymbol{p}_v, \boldsymbol{e})$. When K is negative, $\boldsymbol{e}_u \times \boldsymbol{e}_v$ is a vector with the same direction as $-\boldsymbol{e}$ and $(\boldsymbol{e}_u, \boldsymbol{e}_v, \boldsymbol{e})$ is the left-handed system. Thus when $(\boldsymbol{e}_u, \boldsymbol{e}_v, \boldsymbol{e})$ is the left-handed system, we can define the area of the parallelogram spanned by $\boldsymbol{e}_u$ and $\boldsymbol{e}_v$ as $-|\boldsymbol{e}_u \times \boldsymbol{e}_v|$. Thus, by assigning a (positive or negative) sign to the area, we see from (2.2.53) that K is the ratio of the (signed) area of the two parallelograms. In Figure 2.2.7, we took a small rectangle on

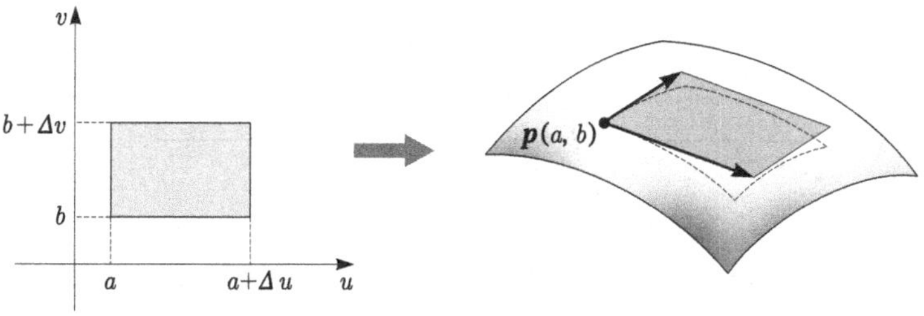

Fig. 2.2.7

the (u, v)-plane and surrounded the corresponding part of the surface by dotted lines. Since its area is almost the same as the area $|\boldsymbol{p}_u \times \boldsymbol{p}_v|\Delta u \Delta v$ of the parallelogram spanned by two vectors $\boldsymbol{p}_u(a, b)\Delta u$ and $\boldsymbol{p}_v(a, b)\Delta v$ on the tangent plane at $\boldsymbol{p}(a, b)$, we call

$$|\boldsymbol{p}_u \times \boldsymbol{p}_v|\, du\, dv \tag{2.2.55}$$

the **area element** of the surface. When (u, v) moves within the region R, the area of the corresponding region on the surface is given by

$$\iint_R |\boldsymbol{p}_u \times \boldsymbol{p}_v| \, du \, dv, \tag{2.2.56}$$

and the signed area of the corresponding region on the unit sphere under the Gauss map is given by

$$\iint_R K \cdot |\boldsymbol{p}_u \times \boldsymbol{p}_v| \, du \, dv. \tag{2.2.57}$$

The area element (2.2.55) of the surface can be written in terms of the first fundamental form as

$$|\boldsymbol{p}_u \times \boldsymbol{p}_v| \, du \, dv = \sqrt{EG - F^2} \, du \, dv. \tag{2.2.58}$$

This can be obtained easily by setting $\boldsymbol{w} = \boldsymbol{y} = \boldsymbol{p}_u$ and $\boldsymbol{x} = \boldsymbol{z} = \boldsymbol{p}_v$ in **Lagrange's formula** for space vectors $\boldsymbol{w}, \boldsymbol{x}, \boldsymbol{y}, \boldsymbol{z}$:

$$(\boldsymbol{w} \times \boldsymbol{x}) \cdot (\boldsymbol{y} \times \boldsymbol{z}) = (\boldsymbol{w} \cdot \boldsymbol{y})(\boldsymbol{x} \cdot \boldsymbol{z}) - (\boldsymbol{w} \cdot \boldsymbol{z})(\boldsymbol{x} \cdot \boldsymbol{y}). \tag{2.2.59}$$

Problem 2.2.1 Given a surface in the form of $z = f(x, y)$, let

$$p = \frac{\partial f}{\partial x}, \quad q = \frac{\partial f}{\partial y}, \quad r = \frac{\partial^2 f}{\partial x^2}, \quad s = \frac{\partial^2 f}{\partial x \partial y}, \quad t = \frac{\partial^2 f}{\partial y^2}.$$

Then prove that the unit normal vector $\boldsymbol{e}$, the first and second fundamental forms, the Gaussian curvature $K = \kappa_1 \kappa_2$, and the mean curvature $H = \frac{\kappa_1 + \kappa_2}{2}$ are given by the following equations.

$$\begin{aligned}
\boldsymbol{e} &= \left(\frac{-p}{\sqrt{1 + p^2 + q^2}}, \frac{-q}{\sqrt{1 + p^2 + q^2}}, \frac{1}{\sqrt{1 + p^2 + q^2}} \right), \\
E &= 1 + p^2, \quad F = pq, \quad G = 1 + q^2, \\
L &= \frac{r}{\sqrt{1 + p^2 + q^2}}, \quad M = \frac{s}{\sqrt{1 + p^2 + q^2}}, \quad N = \frac{t}{\sqrt{1 + p^2 + q^2}}, \\
K &= \frac{rt - s^2}{(1 + p^2 + q^2)^2}, \quad H = \frac{r(1 + q^2) - 2pqs + t(1 + p^2)}{2(1 + p^2 + q^2)^{\frac{3}{2}}}.
\end{aligned}$$

Problem 2.2.2 The **third fundamental form** of a surface $\boldsymbol{p}(u, v)$ is defined by

$$\mathrm{I\!I\!I} = d\boldsymbol{e} \cdot d\boldsymbol{e}.$$

Then, prove the following relationship between the fundamental forms **I**, **II**, and **III**.

$$K\,\mathrm{I} - 2H\,\mathrm{I\!I} + \mathrm{I\!I\!I} = 0.$$

Problem 2.2.3 Prove that the necessary and sufficient condition for a curve $\boldsymbol{p}(s) = \boldsymbol{p}(u(s), v(s))$ on a surface $\boldsymbol{p}(u, v)$ to be a geodesic (i.e., $\boldsymbol{k}_g = \boldsymbol{0}$) is that $(u(s), v(s))$ is a solution to the following system of differential equations,

$$\frac{d^2u}{ds^2} + \Gamma^u_{uu}\frac{du}{ds}\frac{du}{ds} + 2\Gamma^u_{uv}\frac{du}{ds}\frac{dv}{ds} + \Gamma^u_{vv}\frac{dv}{ds}\frac{dv}{ds} = 0,$$
$$\frac{d^2v}{ds^2} + \Gamma^v_{uu}\frac{du}{ds}\frac{du}{ds} + 2\Gamma^v_{uv}\frac{du}{ds}\frac{dv}{ds} + \Gamma^v_{vv}\frac{dv}{ds}\frac{dv}{ds} = 0.$$

Problem 2.2.4 Let $\boldsymbol{e}$ be the unit normal vector of a surface $\boldsymbol{p}(u,v)$ and f a function on the surface. Fix a real number ε with small absolute value $|\varepsilon| > 0$, and consider a new surface $\bar{\boldsymbol{p}} = \boldsymbol{p} + \varepsilon f \boldsymbol{e}$, (i.e., deform the surface $\boldsymbol{p}(u,v)$ in the direction of its normal). Denote by $A(\varepsilon)$ the area $\iint_R |\bar{\boldsymbol{p}}_u \times \bar{\boldsymbol{p}}_v|\, du\, dv$ of $\bar{\boldsymbol{p}}(u,v)$ as (u,v) moves within region R. Prove

$$\left[\frac{dA(\varepsilon)}{d\varepsilon}\right]_{\varepsilon=0} = -\iint_R fH\sqrt{EG-F^2}\, du\, dv.$$

From this equation we see that when $\boldsymbol{p}(u,v)$ is a minimal surface (i.e., $H = 0$), for whatever f we use to deform the surface, we have $A'(0) = 0$. Conversely if $A'(0) = 0$, (in particular, if the area $A(\varepsilon)$ is minimized at $\varepsilon = 0$) for every f, then substituting $f = H$ to the above equation yields

$$0 = \iint_R H^2\sqrt{EG-F^2}\, du\, dv,$$

thus giving $H = 0$. If we want to restrict ourselves to deformations such that $f = 0$ on the boundary of region R, we choose a function φ which is positive inside R and 0 at the boundary. Then by substituting $f = \varphi H$, we get $H = 0$ in the same way. This is why we call a surface with mean curvature $H = 0$ a "minimal surface."

Problem 2.2.5 Prove that a surface is a plane if its second fundamental form II vanishes everywhere.

Problem 2.2.6 We call a surface made by a one-parameter family of lines a **ruled surface** (see the figure in the answer section). It is written as $\boldsymbol{p}(u,v) = \boldsymbol{q}(u) + v\boldsymbol{t}(u)$, where $\boldsymbol{q}(u)$ is a space curve with parameter u, and for each u, $\boldsymbol{t}(u)$ is a nonzero vector. When we fix u and vary v, $\boldsymbol{q}(u) + v\boldsymbol{t}(u)$ gives a line through point $\boldsymbol{q}(u)$ with direction $\boldsymbol{t}(u)$.

(a) For such a surface, prove that $K \leq 0$ everywhere.
(b) Prove that $K \equiv 0$ if and only if $\boldsymbol{t}'(u)$ is written as a linear combination of $\boldsymbol{q}'(u)$ and $\boldsymbol{t}(u)$ such as $\boldsymbol{t}'(u) = \alpha(u)\boldsymbol{q}'(u) + \beta(u)\boldsymbol{t}(u)$. Also, prove that this is equivalent to the geometric condition that the tangent plane at point $\boldsymbol{q}(u) + v\boldsymbol{t}(u)$ does not depend on v.

We call a surface where $K \equiv 0$ and $H \neq 0$ a **developable surface**. The local classification of developable surfaces is classical, but their global classification is a fairly new result. Interested readers should read *Differential Geometry (1969, John Wiley & Sons, Inc.)* by J. J. Stoker.

Problem 2.2.7 (a) Rewrite the hyperboloid of one sheet $\frac{x^2}{a^2} + \frac{y^2}{b^2} - \frac{z^2}{c^2} = 1$ in the form of a ruled surface $\boldsymbol{q}(u) + v\boldsymbol{t}(u)$.
(b) Prove that the hyperbolic paraboloid $z = -\frac{x^2}{a^2} + \frac{y^2}{b^2}$ can also be given in the form of a ruled surface.

2.3 Examples and Calculations of Fundamental Forms and Curvatures

In this section, we shall apply the general theory we learned in the previous section to some simple surfaces.

Example 2.3.1 (Cylindrical surface) A curve

$$x = x(u), \qquad y = y(u), \tag{2.3.1}$$

on the (x, y)-plane, parametrized by u, together with the equation

$$z = v \tag{2.3.2}$$

defines a surface parametrized by (u, v). For simplicity, let the parameter u be normalized so that

$$\left(\frac{dx}{du}\right)^2 + \left(\frac{dy}{du}\right)^2 = 1.$$

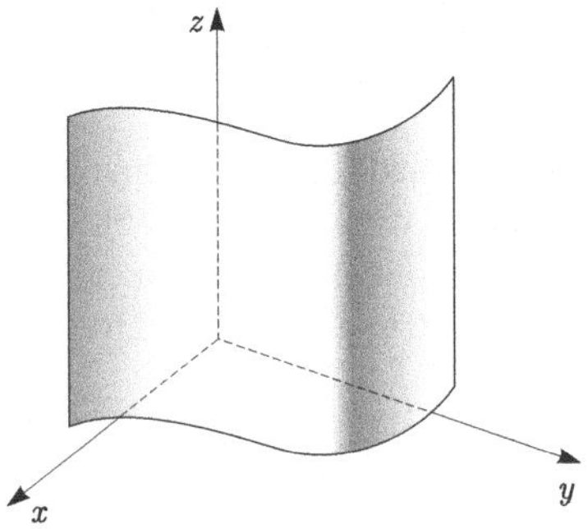

Fig. 2.3.1

With a simple calculation we get,

$$\begin{aligned}
&\boldsymbol{p}(u,v) = (x(u), y(u), v),\\
&d\boldsymbol{p} = \boldsymbol{p}_u\, du + \boldsymbol{p}_v\, dv,\\
&\boldsymbol{p}_u = (x', y', 0), \qquad \boldsymbol{p}_v = (0,0,1),\\
&\mathbf{I} = (du)^2 + (dv)^2,\\
&\text{area element} = du\, dv,\\
&\boldsymbol{e} = (y', -x', 0), \qquad d\boldsymbol{e} = (y''\, du, -x''\, du, 0),\\
&\mathbf{II} = -d\boldsymbol{p}\cdot d\boldsymbol{e} = (x''y' - x'y'')\,(du)^2,
\end{aligned} \tag{2.3.3}$$

so

$$K = 0, \qquad H = \frac{1}{2}(x''y' - x'y''), \tag{2.3.4}$$

and the principal curvatures are

$$\kappa_1 = 0, \qquad \kappa_2 = x''y' - x'y''. \tag{2.3.5}$$

We can easily tell that the principal direction corresponding to κ_1 is the z-axis direction, and the principal direction corresponding to κ_2 is a direction parallel to the (x, y)-plane. κ_2 is -1 times the curvature of the given plane curve (2.3.1). The unit normal $\boldsymbol{e}$ which defines Gauss' spherical map always moves along the equator of the unit sphere, so the area covered by $\boldsymbol{e}$ is obviously zero. This agrees with the fact $K = 0$. ♦

Example 2.3.2 (Ellipsoid : $\frac{x^2}{a^2} + \frac{y^2}{b^2} + \frac{z^2}{c^2} = 1$*)*
Using parameters u and v as in (2.1.16), we get

$$\begin{aligned}
&\boldsymbol{p}(u,v) = (a\cos u\, \cos v,\ b\cos u\, \sin v,\ c\sin u),\\
&\boldsymbol{p}_u = (-a\sin u\, \cos v,\ -b\sin u\, \sin v,\ c\cos u),\\
&\boldsymbol{p}_v = (-a\cos u\, \sin v,\ b\cos u\, \cos v,\ 0),\\
&\boldsymbol{p}_{uu} = (-a\cos u\, \cos v,\ -b\cos u\, \sin v,\ -c\sin u),\\
&\boldsymbol{p}_{uv} = \boldsymbol{p}_{vu} = (a\sin u\, \sin v,\ -b\sin u\, \cos v,\ 0),\\
&\boldsymbol{p}_{vv} = (-a\cos u\, \cos v,\ -b\cos u\, \sin v,\ 0),\\
&\boldsymbol{e} = \frac{1}{\Delta}(-bc\cos u\, \cos v,\ -ca\cos u\, \sin v,\ -ab\sin u),\\
&\qquad \text{where } \Delta = \sqrt{b^2c^2\cos^2 u\cos^2 v + c^2a^2\cos^2 u\sin^2 v + a^2b^2\sin^2 u},\\
&E = a^2\sin^2 u\, \cos^2 v + b^2\sin^2 u\, \sin^2 v + c^2\cos^2 u,\\
&F = (a^2 - b^2)\sin u\, \cos u \sin v\, \cos v,\\
&G = a^2\cos^2 u\, \sin^2 v + b^2\cos^2 u\, \cos^2 v,\\
&EG - F^2 = \Delta^2\cos^2 u,\\
&L = \frac{abc}{\Delta}, \qquad M = 0, \qquad N = \frac{abc\cos^2 u}{\Delta}, \qquad K = \frac{a^2b^2c^2}{\Delta^4},\\
&H = \frac{abc(a^2\sin^2 u\, \cos^2 v + b^2\sin^2 u\, \sin^2 v + c^2\cos^2 u + a^2\sin^2 v + b^2\cos^2 v)}{2\Delta^3}\\
&\quad = \frac{abc[(a^2 + b^2 + c^2) - (a^2\cos^2 u\, \cos^2 v + b^2\cos^2 u\, \sin^2 v + c^2\sin^2 u)]}{2\Delta^3}.
\end{aligned} \tag{2.3.6}$$

Using (x, y, z) we have

$$K = \frac{1}{a^2b^2c^2\left(\frac{x^2}{a^4} + \frac{y^2}{b^4} + \frac{z^2}{c^4}\right)^2}, \quad H = \frac{(a^2+b^2+c^2) - (x^2+y^2+z^2)}{2a^2b^2c^2\left(\frac{x^2}{a^4} + \frac{y^2}{b^4} + \frac{z^2}{c^4}\right)^{\frac{3}{2}}}.$$

In particular, for a sphere with radius a (i.e., $a = b = c$), we have,

$$\begin{aligned}
&\mathbf{I} = a^2(du)^2 + a^2\cos^2 u\,(dv)^2,\\
&\mathbf{II} = a(du)^2 + a\cos^2 u\,(dv)^2,\\
&\boldsymbol{e} = \left(-\frac{x}{a}, -\frac{y}{a}, -\frac{z}{a}\right),\\
&K = \frac{1}{a^2}, \qquad H = \frac{1}{a}.
\end{aligned} \tag{2.3.7}$$

♦

Example 2.3.3 (Hyperboloid of one sheet : $\frac{x^2}{a^2} + \frac{y^2}{b^2} - \frac{z^2}{c^2} = 1$)

Using equation (2.1.17), as in the calculation of an ellipsoid we have

$$\begin{aligned}
&\boldsymbol{p}(u, v) = (a\cosh u\ \cos v,\ b\cosh u\ \sin v,\ c\sinh u),\\
&\boldsymbol{e} = \frac{-1}{\left(\frac{x^2}{a^4} + \frac{y^2}{b^4} + \frac{z^2}{c^4}\right)^{\frac{1}{2}}}\left(\frac{x}{a^2}, \frac{y}{b^2}, -\frac{z}{c^2}\right),\\
&E = a^2\sinh^2 u\ \cos^2 v + b^2\sinh^2 u\ \sin^2 v + c^2\cosh^2 u,\\
&F = (b^2 - a^2)\sinh u\ \cosh u\ \sin v\ \cos v,\\
&G = a^2\cosh^2 u\ \sin^2 v + b^2\cosh^2 u\ \cos^2 v,\\
&L = \frac{-1}{\left(\frac{x^2}{a^4} + \frac{y^2}{b^4} + \frac{z^2}{c^4}\right)^{\frac{1}{2}}}, \quad M = 0, \quad N = \frac{\cosh^2 u}{\left(\frac{x^2}{a^4} + \frac{y^2}{b^4} + \frac{z^2}{c^4}\right)^{\frac{1}{2}}},\\
&K = \frac{-1}{a^2b^2c^2\left(\frac{x^2}{a^4} + \frac{y^2}{b^4} + \frac{z^2}{c^4}\right)^2},\\
&H = \frac{(x^2+y^2+z^2) - (a^2+b^2-c^2)}{2a^2b^2c^2\left(\frac{x^2}{a^4} + \frac{y^2}{b^4} + \frac{z^2}{c^4}\right)^{\frac{3}{2}}}.
\end{aligned} \tag{2.3.8}$$

♦

Example 2.3.4 (Hyperboloid of two sheets : $\frac{x^2}{a^2}+\frac{y^2}{b^2}-\frac{z^2}{c^2}=-1$) From equation (2.1.18) we have

$$
\begin{aligned}
&\boldsymbol{p}(u,v)=(a\sinh u\,\cos v,\; b\sinh u\,\sin v,\; c\cosh u),\\
&\boldsymbol{e}=\frac{-1}{\left(\frac{x^2}{a^4}+\frac{y^2}{b^4}+\frac{z^2}{c^4}\right)^{\frac{1}{2}}}\left(\frac{x}{a^2},\frac{y}{b^2},-\frac{z}{c^2}\right),\\
&E=a^2\cosh^2 u\,\cos^2 v+b^2\cosh^2 u\,\sin^2 v+c^2\sinh^2 u,\\
&F=(b^2-a^2)\sinh u\,\cosh u\,\sin v\,\cos v,\\
&G=a^2\sinh^2 u\,\sin^2 v+b^2\sinh^2 u\,\cos^2 v,\\
&L=\frac{1}{\left(\frac{x^2}{a^4}+\frac{y^2}{b^4}+\frac{z^2}{c^4}\right)^{\frac{1}{2}}},\quad M=0,\quad N=\frac{\sinh^2 u}{\left(\frac{x^2}{a^4}+\frac{y^2}{b^4}+\frac{z^2}{c^4}\right)^{\frac{1}{2}}},\\
&K=\frac{1}{a^2b^2c^2\left(\frac{x^2}{a^4}+\frac{y^2}{b^4}+\frac{z^2}{c^4}\right)^2},\\
&H=\frac{(x^2+y^2+z^2)+(a^2+b^2-c^2)}{2a^2b^2c^2\left(\frac{x^2}{a^4}+\frac{y^2}{b^4}+\frac{z^2}{c^4}\right)^{\frac{3}{2}}}.
\end{aligned}
\tag{2.3.9}
$$

♦

Example 2.3.5 (Torus) From equation (2.1.21) we obtain

$$
\begin{aligned}
&\boldsymbol{p}(u,v)=((R+r\cos u)\cos v,\;(R+r\cos u)\sin v,\;r\sin u),\\
&\boldsymbol{e}=(-\cos u\,\cos v,\;-\cos u\,\sin v,\;-\sin u),\\
&E=r^2,\quad F=0,\quad G=(R+r\cos u)^2,\\
&L=r,\quad M=0,\quad N=(R+r\cos u)\cos u,\\
&K=\frac{\cos u}{r(R+r\cos u)},\quad 2H=\frac{\cos u}{R+r\cos u}+\frac{1}{r},\\
&\kappa_1=\frac{\cos u}{R+r\cos u},\quad \kappa_2=\frac{1}{r},
\end{aligned}
\tag{2.3.10}
$$

where the principal direction corresponding to κ_1 is horizontal in Figure 2.1.8 and the principal direction corresponding to κ_2 is perpendicular to it. From the sign of $\cos u$, we see that $K = 0$ along the circle at the top as well as the circle at the bottom and that $K > 0$ in the outer region surrounded by these two circles and $K < 0$ in the inner region. This fact agrees with Theorem 2.2.2. ◆

More generally we consider

Example 2.3.6 (Surface of revolution)

Using equation (2.1.24) and with calculations similar to Example 2.3.5, we have

$\boldsymbol{p}(u, v) = (f(u)\cos v,\ f(u)\sin v,\ g(u))$ (Since the curve $x = f(u), z = g(u)$ does not intersect the z-axis, we may assume $f(u) > 0$.)

$$\boldsymbol{e} = \frac{1}{\sqrt{f'(u)^2 + g'(u)^2}}(-g'(u)\cos v,\ -g'(u)\sin v,\ f'(u)),$$

$$E = f'(u)^2 + g'(u)^2, \quad F = 0, \quad G = f^2,$$

$$L = \frac{f'(u)g''(u) - f''(u)g'(u)}{\sqrt{f'(u)^2 + g'(u)^2}}, \quad M = 0,$$

$$N = \frac{f(u)g'(u)}{\sqrt{f'(u)^2 + g'(u)^2}},$$

$$K = \frac{\{f'(u)g''(u) - f''(u)g'(u)\}g'(u)}{f(u)\{f'(u)^2 + g'(u)^2\}^2},$$

$$2H = \frac{g'(u)}{f(u)\{f'(u)^2 + g'(u)^2\}^{\frac{1}{2}}} + \frac{f'(u)g''(u) - f''(u)g'(u)}{\{f'(u)^2 + g'(u)^2\}^{\frac{3}{2}}},$$

$$\kappa_1 = \frac{g'(u)}{f(u)\{f'(u)^2 + g'(u)^2\}^{\frac{1}{2}}},$$

$$\kappa_2 = \frac{f'(u)g''(u) - f''(u)g'(u)}{\{f'(u)^2 + g'(u)^2\}^{\frac{3}{2}}}. \tag{2.3.11}$$

When expressing the curve on the (x, z)-plane, we can either write $x = f(u), z = g(u)$ choosing a parameter u such that

$$f'(u)^2 + g'(u)^2 = 1, \tag{2.3.12}$$

or write

$$x = f(z) \quad \text{or} \quad z = g(x) \tag{2.3.13}$$

to simplify the equations for K, H, κ_1, κ_2.

For example, choose a parameter u as in (2.3.12). Using (2.3.12) and its differentiation

$$f'(u)f''(u) + g'(u)g''(u) = 0, \tag{2.3.14}$$

we obtain

$$(f'g'' - f''g')g' = f'g''g' - f''g'g' = -f'f'f'' - f''(1 - f'f') = -f''. \tag{2.3.15}$$

Thus we have

$$\begin{aligned} K &= -\frac{f''}{f}, \quad 2H = \frac{g'}{f} - \frac{f''}{g'}, \\ \kappa_1 &= \frac{g'}{f}, \qquad \kappa_2 = -\frac{f''}{g'}. \end{aligned} \tag{2.3.16}$$

♦

We already know from Example 2.3.2 that a sphere of radius a has curvature $K = \frac{1}{a^2}$. Now using (2.3.16) we construct a surface whose curvature K is a negative constant.

Example 2.3.7 (Surface of revolution with curvature $K = -c^2$) In (2.3.16) let $K = -c^2$ $(c > 0)$. Then we have

$$f''(u) = c^2 f(u). \tag{2.3.17}$$

Taking

$$f(u) = \frac{1}{c}e^{-cu} \tag{2.3.18}$$

as a solution to this differential equation, we obtain

$$\begin{aligned} g'(u) &= \pm\sqrt{1 - f'(u)^2} = \pm\sqrt{1 - e^{-2cu}}, \\ g(u) &= \pm\int_0^u \sqrt{1 - e^{-2ct}}\,dt \quad (0 \le u < \infty). \end{aligned} \tag{2.3.19}$$

See Figure 2.3.2 for the graph of the curve $x = f(u)$, $z = g(u)$. Revolving it around

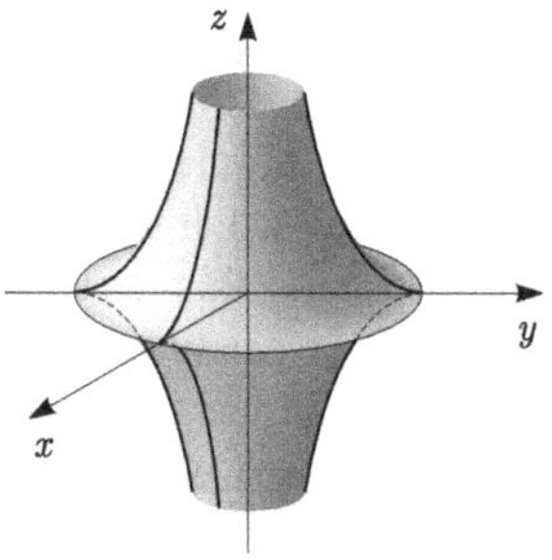

Fig. 2.3.2

the z-axis, we get a surface with curvature $-c^2$, which, however, is not smooth where it intersects the (x, y)-plane. This surface, called a **pseudosphere**, is important since it realizes non-Euclidean geometry locally in the three-dimensional Euclidean space. ♦

Next using (2.3.16) we construct a surface of revolution with $H = 0$.

Example 2.3.8 (Catenoid) Let $H = 0$ in (2.3.16). Then we have

$$ff'' = (g')^2 = 1 - (f')^2. \tag{2.3.20}$$

Thus

$$(ff')' = (f')^2 + ff'' = 1. \tag{2.3.21}$$

Integrating this we get

$$\frac{1}{2}(f^2)' = ff' = u + c, \tag{2.3.22}$$

where c is a constant. Integrating it once more we get

$$\begin{aligned} f^2 &= u^2 + 2cu + d, \\ f &= \sqrt{u^2 + 2cu + d}, \end{aligned} \tag{2.3.23}$$

where d is a constant. Thus

$$g' = \pm\sqrt{1 - (f')^2} = \pm\sqrt{\frac{d - c^2}{u^2 + 2cu + d}}. \tag{2.3.24}$$

For simplicity let $c = 0$, $d = a^2$ $(a > 0)$. Then

$$\begin{aligned} x &= f(u) = \sqrt{u^2 + a^2}, \\ z &= g(u) = \pm \int_0^u \frac{a\,dt}{\sqrt{t^2 + a^2}} = \pm a \sinh^{-1} \frac{u}{a}. \end{aligned} \tag{2.3.25}$$

In order to draw the graph of this curve, eliminate u from (2.3.25) to get

$$x = a \cosh \frac{z}{a}, \tag{2.3.26}$$

and this curve is called a **catenary**. See Figure 2.3.3 for its graph. If we rotate the graph by 90° so that the x axis points upward, it will look like a chain hanging freely between two points of support. Revolving this curve around the z-axis gives us a catenoid. We assumed $c = 0$ in the above calculation for simplicity. Since replacing the parameter u by $u - c$ eliminate the linear term, this assumption can be made without loss of generality. In order for $f(u) = \sqrt{u^2 + d}$ to be defined for all u and not to intersect the z-axis, we assumed $d > 0$. Thus we let $d = a^2$. A catenoid is an important example of a minimal surface.

♦

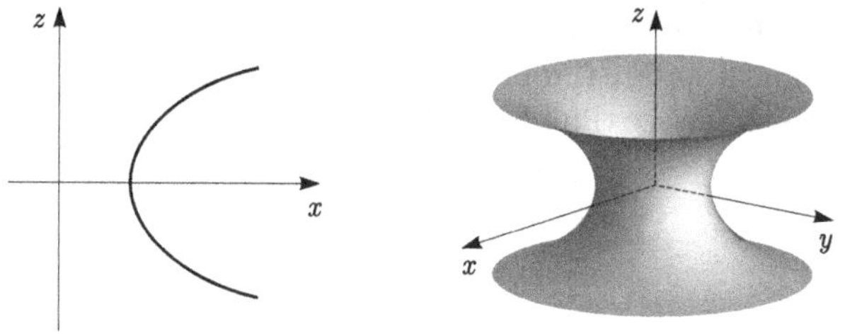

Fig. 2.3.3

Problem 2.3.1 Calculate the Gaussian curvature K and the mean curvature H for an elliptic paraboloid and a hyperbolic paraboloid. (Calculated using the parametric representations in Problem 2.1.1 as well as those in Examples 2.1.5 and 2.1.6.)

Problem 2.3.2 We call a surface of the form

$$x = u\cos v, \quad y = u\sin v, \quad z = f(v)$$

a **right conoid**, (see Figure 5.2.1 in Chapter 5). This is a ruled surface formed by lines perpendicularly intersecting the z-axis. Calculate the curvatures K and H of this surface. When $f(v) = av + b$, this conoid is called a **right helicoid** (see Figure 5.2.2 in Chapter 5). Check that a right helicoid is a minimal surface and that a right conoid which is a minimal surface is a right helicoid.

Problem 2.3.3 A surface given by

$$x = 3u + 3uv^2 - u^3, \quad y = v^3 - 3v - 3u^2v, \quad z = 3(u^2 - v^2)$$

is called the **Enneper's minimal surface** (see Figure 5.4.1 in Chapter 5). Check that this is a minimal surface and calculate its Gaussian curvature K.

Problem 2.3.4 A surface defined by $e^z\cos x = \cos y$ is called the **Scherk's minimal surface** (see Figure 5.2.4 in Chapter 5). Check that this is a minimal surface and calculate its Gaussian curvature K.

Problem 2.3.5 Find the umbilic points of an ellipsoid $\frac{x^2}{a^2} + \frac{y^2}{b^2} + \frac{z^2}{c^2} = 1$ $(a > b > c > 0)$.

Problem 2.3.6 For each of the following surfaces determine which part of the sphere is covered by Gauss' spherical map.

(a) Ellipsoid (Examples 2.1.2, 2.3.2).
(b) Hyperboloid of one sheet (Examples 2.1.3, 2.3.3).
(c) Hyperboloid of two sheets (Examples 2.1.4, 2.3.4).
(d) Elliptic paraboloid (Examples 2.1.5, Problem 2.3.1).
(e) Hyperbolic paraboloid (Examples 2.1.6, Problem 2.3.1).
(f) Torus (Examples 2.1.7, 2.3.5).
(g) Catenoid (Example 2.3.8).

(h) Right helicoid (Problem 2.3.2).
(i) Enneper's minimal surface (Problem 2.3.3).
(j) Scherk's minimal surface (Problem 2.3.4).

2.4 Method of Orthonormal Frames

Having seen Frenet-Serret's formula (1.4.14), we can easily expect that various formulas for a surface $\boldsymbol{p}(u, v)$ would be more beautifully expressed in terms of unit vectors orthogonal to each other (**orthonormal frame**), than in terms of vectors $\boldsymbol{p}_u$, $\boldsymbol{p}_v$, and $\boldsymbol{e}$. In this section, we shall review what we have calculated in Section 2.2 using orthonormal frames.

Given a surface $\boldsymbol{p}(u, v)$, at each point we choose vectors $\boldsymbol{e}_1$ and $\boldsymbol{e}_2$ in its tangent plane so that we have

$$\boldsymbol{e}_1 \cdot \boldsymbol{e}_1 = \boldsymbol{e}_2 \cdot \boldsymbol{e}_2 = 1, \qquad \boldsymbol{e}_1 \cdot \boldsymbol{e}_2 = 0. \tag{2.4.1}$$

In this case, we choose $\boldsymbol{e}_1$ and $\boldsymbol{e}_2$ in such a way that they can be differentiated a number of times in (u, v). We can see that this is possible by setting

$$\boldsymbol{e}_1 = \frac{\boldsymbol{p}_u}{|\boldsymbol{p}_u|}, \qquad \boldsymbol{e}_2 = \frac{\boldsymbol{p}_v - (\boldsymbol{p}_v \cdot \boldsymbol{e}_1)\boldsymbol{e}_1}{|\boldsymbol{p}_v - (\boldsymbol{p}_v \cdot \boldsymbol{e}_1)\boldsymbol{e}_1|}, \tag{2.4.2}$$

but the frame $\boldsymbol{e}_1$, $\boldsymbol{e}_2$ we use below need not be defined as in (2.4.2). Next we set

$$\boldsymbol{e}_3 = \boldsymbol{e}_1 \times \boldsymbol{e}_2. \tag{2.4.3}$$

Since $\boldsymbol{e}_3$ is a unit vector orthogonal to $\boldsymbol{e}_1$ and $\boldsymbol{e}_2$, it is a unit normal vector, and it is equal to the vector $\boldsymbol{e} = \frac{\boldsymbol{p}_u \times \boldsymbol{p}_v}{|\boldsymbol{p}_u \times \boldsymbol{p}_v|}$ up to a sign. If $\boldsymbol{e}_3 = -\boldsymbol{e}$, by interchanging $\boldsymbol{e}_1$ and $\boldsymbol{e}_2$ we get $\boldsymbol{e}_3 = \boldsymbol{e}$. So we assume that $\boldsymbol{e}_3 = \boldsymbol{e}$. Since both $\boldsymbol{e}_1$, $\boldsymbol{e}_2$ and $\boldsymbol{p}_u$, $\boldsymbol{p}_v$ are bases for each tangent plane, we get

$$\begin{aligned} \boldsymbol{p}_u &= {a_1}^1 \boldsymbol{e}_1 + {a_1}^2 \boldsymbol{e}_2, \\ \boldsymbol{p}_v &= {a_2}^1 \boldsymbol{e}_1 + {a_2}^2 \boldsymbol{e}_2, \end{aligned} \qquad (\boldsymbol{p}_u, \boldsymbol{p}_v) = (\boldsymbol{e}_1, \boldsymbol{e}_2) \begin{bmatrix} {a_1}^1 & {a_2}^1 \\ {a_1}^2 & {a_2}^2 \end{bmatrix}, \tag{2.4.4}$$

and the matrix

$$A = \begin{bmatrix} {a_1}^1 & {a_2}^1 \\ {a_1}^2 & {a_2}^2 \end{bmatrix} \tag{2.4.5}$$

has nonzero determinant $\det A = {a_1}^1 {a_2}^2 - {a_2}^1 {a_1}^2$. Since we have chosen $\boldsymbol{e}_1$ and $\boldsymbol{e}_2$ such that $\boldsymbol{e}_3 = \boldsymbol{e}$, we always have

$$\det A > 0. \tag{2.4.6}$$

In fact, by simple calculations we can check that

$$\boldsymbol{p}_u \times \boldsymbol{p}_v = (\det A)\boldsymbol{e}_3. \tag{2.4.7}$$

Since

$$\begin{aligned} d\boldsymbol{p} &= \boldsymbol{p}_u\, du + \boldsymbol{p}_v\, dv \\ &= ({a_1}^1\, du + {a_2}^1\, dv)\, \boldsymbol{e}_1 + ({a_1}^2\, du + {a_2}^2\, dv)\, \boldsymbol{e}_2, \end{aligned} \tag{2.4.8}$$

by setting

$$\theta^1 = {a_1}^1\, du + {a_2}^1\, dv, \quad \theta^2 = {a_1}^2\, du + {a_2}^2\, dv, \tag{2.4.9}$$

we get

$$d\boldsymbol{p} = \theta^1\, \boldsymbol{e}_1 + \theta^2\, \boldsymbol{e}_2. \tag{2.4.10}$$

Thus the first fundamental form **I** will be given by

$$\mathbf{I} = d\boldsymbol{p} \cdot d\boldsymbol{p} = \theta^1\theta^1 + \theta^2\theta^2. \tag{2.4.11}$$

Since $\boldsymbol{e}_1$, $\boldsymbol{e}_2$, and $\boldsymbol{e}_3$ form an orthonormal frame, in particular, a basis of the three-dimensional vector space, $d\boldsymbol{e}_1$, $d\boldsymbol{e}_2$, and $d\boldsymbol{e}_3$ can be written as linear combinations of $\boldsymbol{e}_1$, $\boldsymbol{e}_2$ and $\boldsymbol{e}_3$. Thus

$$\begin{aligned} d\boldsymbol{e}_1 &= {\omega_1}^1\, \boldsymbol{e}_1 + {\omega_1}^2\, \boldsymbol{e}_2 + {\omega_1}^3\, \boldsymbol{e}_3, \\ d\boldsymbol{e}_2 &= {\omega_2}^1\, \boldsymbol{e}_1 + {\omega_2}^2\, \boldsymbol{e}_2 + {\omega_2}^3\, \boldsymbol{e}_3, \\ d\boldsymbol{e}_3 &= {\omega_3}^1\, \boldsymbol{e}_1 + {\omega_3}^2\, \boldsymbol{e}_2 + {\omega_3}^3\, \boldsymbol{e}_3, \end{aligned} \tag{2.4.12}$$

where ${\omega_j}^i$ is a linear combination of du and dv. (We can explicitly write out ${\omega_j}^i$ using du and dv, but for now it is not necessary to do so.) Since $\boldsymbol{e}_i \cdot \boldsymbol{e}_j$ is a constant, by differentiating it we get

$$d\boldsymbol{e}_i \cdot \boldsymbol{e}_j + \boldsymbol{e}_i \cdot d\boldsymbol{e}_j = 0 \qquad (i, j = 1, 2, 3), \tag{2.4.13}$$

and by substituting (2.4.12) into this, we obtain

$${\omega_i}^j + {\omega_j}^i = 0 \qquad (i, j = 1, 2, 3). \tag{2.4.14}$$

Thus the matrix made from ${\omega_j}^i$ in (2.4.12) is skew-symmetric. In particular, ${\omega_1}^1 = {\omega_2}^2 = {\omega_3}^3 = 0$.

From the definition (2.2.20), we can write the second fundamental form $\mathbf{II}$ as follows:

$$\begin{aligned} \mathbf{II} = -d\boldsymbol{p} \cdot d\boldsymbol{e}_3 &= -(\theta^1\, \boldsymbol{e}_1 + \theta^2\, \boldsymbol{e}_2) \cdot ({\omega_3}^1\, \boldsymbol{e}_1 + {\omega_3}^2\, \boldsymbol{e}_2) \\ &= -\theta^1\, {\omega_3}^1 - \theta^2\, {\omega_3}^2 = \theta^1\, {\omega_1}^3 + \theta^2\, {\omega_2}^3. \end{aligned} \tag{2.4.15}$$

Since the matrix A in (2.4.5) is an invertible matrix, du and dv can be written as a linear combination of θ^1 and θ^2. Since ${\omega_1}^3$ and ${\omega_2}^3$ are linear combinations of du and dv, they can be written as linear combinations of θ^1 and θ^2. Therefore, we can set

$$\begin{aligned} {\omega_1}^3 &= b_{11}\, \theta^1 + b_{12}\, \theta^2, \\ {\omega_2}^3 &= b_{21}\, \theta^1 + b_{22}\, \theta^2. \end{aligned} \tag{2.4.16}$$

Thus, from (2.4.15) and (2.4.16) we have

$$\mathbf{II} = \sum_{i,j=1}^{2} b_{ij}\, \theta^i\, \theta^j. \tag{2.4.17}$$

We prove that the matrix

$$B = \begin{bmatrix} b_{11} & b_{12} \\ b_{21} & b_{22} \end{bmatrix} \tag{2.4.18}$$

is symmetric. By using exterior differential forms this can be proved more easily (as we will see later), but for now let us prove this by comparing B with the symmetric matrix

$$S = \begin{bmatrix} L & M \\ M & N \end{bmatrix}. \tag{2.4.19}$$

Using $d\boldsymbol{e}_3 = d\boldsymbol{e} = \boldsymbol{e}_u\, du + \boldsymbol{e}_v\, dv$, (2.2.23), (2.4.4), (2.4.12) and (2.4.14) we have

$$\begin{aligned} -L\, du - M\, dv &= d\boldsymbol{e}_3 \cdot \boldsymbol{p}_u = ({\omega_3}^1\, \boldsymbol{e}_1 + \omega_3^2\, \boldsymbol{e}_2) \cdot ({a_1}^1\, \boldsymbol{e}_1 + a_1^2\, \boldsymbol{e}_2) \\ &= {\omega_3}^1\, a_1^1 + {\omega_3}^2\, {a_1}^2 = -{\omega_1}^3\, {a_1}^1 - {\omega_2}^3\, {a_1}^2. \end{aligned}$$

Similarly,

$$-M\, du - N\, dv = -{\omega_1}^3\, {a_2}^1 - {\omega_2}^3\, {a_2}^2.$$

Therefore we have

$$\begin{aligned} \begin{bmatrix} L & M \\ M & N \end{bmatrix} \begin{bmatrix} du \\ dv \end{bmatrix} &= \begin{bmatrix} {a_1}^1 & {a_1}^2 \\ {a_2}^1 & {a_2}^2 \end{bmatrix} \begin{bmatrix} {\omega_1}^3 \\ {\omega_2}^3 \end{bmatrix} \\ &= \begin{bmatrix} {a_1}^1 & {a_1}^2 \\ {a_2}^1 & {a_2}^2 \end{bmatrix} \begin{bmatrix} b_{11} & b_{12} \\ b_{21} & b_{22} \end{bmatrix} \begin{bmatrix} \theta^1 \\ \theta^2 \end{bmatrix} \\ &= \begin{bmatrix} {a_1}^1 & {a_1}^2 \\ {a_2}^1 & {a_2}^2 \end{bmatrix} \begin{bmatrix} b_{11} & b_{12} \\ b_{21} & b_{22} \end{bmatrix} \begin{bmatrix} {a_1}^1 & {a_2}^1 \\ {a_1}^2 & {a_2}^2 \end{bmatrix} \begin{bmatrix} du \\ dv \end{bmatrix}, \end{aligned} \tag{2.4.20}$$

in matrix notation,

$$S = {}^t\!ABA, \tag{2.4.21}$$

where ${}^t\!A$ stands for the **transpose** of A. In (2.4.21), since S is symmetric, we see that B is also symmetric.

The eigenvalues of matrix B are real numbers. We write them as κ_1 and κ_2, and call them **principal curvatures**. (We will soon see that they are equal to the principal curvatures defined in Section 2.2.) The Gaussian curvature K and the mean curvature H are defined to be

$$\begin{aligned} K &= \kappa_1 \kappa_2, \\ H &= \frac{1}{2}(\kappa_1 + \kappa_2). \end{aligned} \tag{2.4.22}$$

Using the orthogonal matrix P we diagonalize B as

$${}^t\!PBP = \begin{bmatrix} \kappa_1 & 0 \\ 0 & \kappa_2 \end{bmatrix} \tag{2.4.23}$$

to see that

$$\begin{aligned} K &= \det B = b_{11}\, b_{22} - b_{12}\, b_{21}, \\ 2H &= \operatorname{trace} B = b_{11} + b_{22}. \end{aligned} \tag{2.4.24}$$

On the other hand, in Section 2.2 we defined principal curvatures κ_1 and κ_2 as solutions to (2.2.42). To prove that they are equal to the κ_1 and κ_2 which were defined in Section 2.2 as eigenvalues of B, we have only to check that the Gaussian curvature K and the mean curvature H defined in Section 2.2 are equal to the K and H defined in this section. (Because κ_1 and κ_2 are determined by K and H as solutions to $\lambda^2 - 2H\lambda + K = 0$.)

From (2.4.21) we have

$$B = {}^t A^{-1} S A^{-1} = A A^{-1} \, {}^t A^{-1} S A^{-1}. \tag{2.4.25}$$

Thus,

$$\begin{aligned} \det B &= \det(A^{-1} \, {}^t A^{-1} S) = (\det {}^t A A)^{-1} (\det S), \\ \operatorname{trace} B &= \operatorname{trace}\, (({}^t A A)^{-1} S). \end{aligned} \tag{2.4.26}$$

On the other hand, from (2.4.4) we get

$$\begin{aligned} {}^t A A &= \begin{bmatrix} a_1{}^1 & a_1{}^2 \\ a_2{}^1 & a_2{}^2 \end{bmatrix} \begin{bmatrix} a_1{}^1 & a_2{}^1 \\ a_1{}^2 & a_2{}^2 \end{bmatrix} \\ &= \begin{bmatrix} E & F \\ F & G \end{bmatrix}. \end{aligned} \tag{2.4.27}$$

We can easily calculate the inverse of a 2×2 matrix:

$$\begin{bmatrix} E & F \\ F & G \end{bmatrix}^{-1} = \frac{1}{EG - F^2} \begin{bmatrix} G & -F \\ -F & E \end{bmatrix}. \tag{2.4.28}$$

From (2.4.26), (2.4.27), and (2.4.28) we have

$$\begin{aligned} \det B &= \frac{LN - M^2}{EG - F^2}, \\ \operatorname{trace} B &= \frac{EN + GL - 2FM}{EG - F^2}. \end{aligned} \tag{2.4.29}$$

By comparing this with (2.2.43) we can see that the two definitions of K and H are equal.

From the expressions (2.4.24) for K and H, we see that using orthonormal frames is more natural in developing the general theory. On the other hand, when calculating K and H for concrete examples, it might be more convenient to use as in Section 2.3 the method we have learned in Section 2.2. Before we go further, it would be better to explain exterior differential forms.

Problem 2.4.1 We check the relationship between the matrix A defined in (2.4.5), the matrix ω with entries $\omega_j{}^i$ $(1 \leq i, j \leq 2)$ defined in (2.4.12), and Christoffel's symbols defined in (2.2.25). Here let us use the symbols for tensor analysis; we write u^1, u^2 instead of u, v and $\Gamma^1_{11}, \Gamma^1_{12}, \ldots$ instead of $\Gamma^u_{uu}, \Gamma^u_{uv}, \ldots$ (replacing the indices u, v by $1, 2$). Then prove that the relationship

$$\omega A = A\Gamma - dA \quad (\text{or } \omega = A\Gamma A^{-1} - dA \cdot A^{-1})$$

holds between matrices

$$A = \begin{bmatrix} a_1{}^1 & a_2{}^1 \\ a_1{}^2 & a_2{}^2 \end{bmatrix}, \quad \omega = \begin{bmatrix} \omega_1{}^1 & \omega_2{}^1 \\ \omega_1{}^2 & \omega_2{}^2 \end{bmatrix}, \quad \Gamma = \begin{bmatrix} \gamma_1{}^1 & \gamma_2{}^1 \\ \gamma_1{}^2 & \gamma_2{}^2 \end{bmatrix}, \quad \text{where } \gamma_j{}^i = \sum_{k=1}^{2} \Gamma^i_{jk}\, du^k.$$

Using the components of matrices we have

$$\sum_{k=1}^{2} \omega_k{}^i\, a_j{}^k = \sum_{k,l=1}^{2} a_k{}^i\, \Gamma^k_{jl}\, du^l - da_j{}^i.$$

2.5 Exterior Differential Forms in Two Variables

The expression $adu + bdv$ have already come up in calculus and also in previous sections of this book. This is a special case of an **exterior differential form**, or just simply **differential form**. In this section we will explain exterior differential forms in two variables, and in the next section we will apply results to surfaces.

Differential forms defined on a region D in the (u, v)-plane are made by adding and multiplying functions on D and differentials du, dv. However as we will see below, the multiplication rule is not commutative. We use a special symbol $\wedge$ for multiplication and call it the **exterior multiplication**. It obeys the following rule:

$$du \wedge du = 0, \quad dv \wedge dv = 0, \quad du \wedge dv = -dv \wedge du. \tag{2.5.1}$$

Explicitly,

functions are 0 degree differential forms (0-forms),
$f\,du + g\,dv$'s are linear differential forms (1-forms),
$f\,du \wedge dv$'s are quadratic differential forms (2-forms).

We have already been using 0-forms and 1-forms. Because of (2.5.1), a 2-form $f\,du \wedge dv$ can be written in many different ways, e.g.,

$$f\,du \wedge dv = -f\,dv \wedge du = \frac{1}{2}(f\,du \wedge dv - f\,dv \wedge du).$$

Now, by multiplying two 1-forms

$$\alpha = a_1\,du + a_2\,dv, \qquad \beta = b_1\,du + b_2\,dv,$$

we get

$$\begin{aligned} \alpha \wedge \beta &= (a_1\,du + a_2\,dv) \wedge (b_1\,du + b_2\,dv) \\ &= a_1 b_2\,du \wedge dv + a_2 b_1\,dv \wedge du \\ &= (a_1 b_2 - a_2 b_1)\,du \wedge dv, \end{aligned}$$

where the coefficient $a_1 b_2 - a_2 b_1$ is the determinant of the 2×2 matrix made from coefficients of α and β:

$$\begin{bmatrix} a_1 & a_2 \\ b_1 & b_2 \end{bmatrix}.$$

From this,

$$\alpha \wedge \beta = -\beta \wedge \alpha$$

is obvious, and we see that the condition $\alpha \wedge \beta \neq 0$ is equivalent to the condition that α and β are linearly independent.

From (2.5.1), note that differential forms of degree 3 or higher will be 0. In the case of three variables, we have to consider also 3-forms.

Next we shall define the **exterior differentiation** d. For a 0-form (function) f, we define

$$df = \frac{\partial f}{\partial u}\, du + \frac{\partial f}{\partial v}\, dv. \tag{2.5.2}$$

For a 1-form $\varphi = f\, du + g\, dv$, we define

$$d\varphi = df \wedge du + dg \wedge dv = \left(-\frac{\partial f}{\partial v} + \frac{\partial g}{\partial u}\right) du \wedge dv. \tag{2.5.3}$$

For a 2-form $\psi = f\, du \wedge dv$, we define

$$\begin{aligned} d\psi &= df \wedge du \wedge dv \\ &= \left(\frac{\partial f}{\partial u}\, du + \frac{\partial f}{\partial v}\, dv\right) \wedge du \wedge dv = 0. \end{aligned} \tag{2.5.4}$$

The exterior differentiation d increases the degree by 1. Proofs of the following formulas are easy, so will be left as problems.
If f, g are both functions, then

$$d(fg) = df \cdot g + f \cdot dg. \tag{2.5.5}$$

If f is a function and φ is a 1-form, then

$$d(f\varphi) = df \wedge \varphi + f d\varphi, \tag{2.5.6}$$

$$d(\varphi f) = d\varphi \cdot f - \varphi \wedge df. \tag{2.5.7}$$

When multiplying a function and a differential form, instead of $df \wedge g$ we may write $df \cdot g$. Note in (2.5.7) that the right hand side is not $d\varphi \cdot f + \varphi \wedge df$. This is because df and φ being both of degree 1 so we have $df \wedge \varphi = -\varphi \wedge df$, and (2.5.6) yields (2.5.7). Another important property of d is that for all forms, we have

$$dd\theta = 0. \tag{2.5.8}$$

When θ is of degree 1 or 2, $dd\theta$ is of degree 3 or 4, so it is obviously 0. When θ is a function, we can check this as follows:

$$\begin{aligned} ddf &= d\left(\frac{\partial f}{\partial u}\,du + \frac{\partial f}{\partial v}\,dv\right) \\ &= \left(\frac{\partial^2 f}{\partial u \partial u}\,du + \frac{\partial^2 f}{\partial u \partial v} dv\right) \wedge du + \left(\frac{\partial^2 f}{\partial v \partial u}\,du + \frac{\partial^2 f}{\partial v \partial v}\,dv\right) \wedge dv \\ &= \frac{\partial^2 f}{\partial u \partial v}\,dv \wedge du + \frac{\partial^2 f}{\partial v \partial u}\,du \wedge dv \\ &= \left(-\frac{\partial^2 f}{\partial u \partial v} + \frac{\partial^2 f}{\partial v \partial u}\right) du \wedge dv = 0. \end{aligned} \tag{2.5.9}$$

Now, we consider the converse of (2.5.8). "If a 1-form $\varphi = f du + g dv$ satisfies $d\varphi = 0$, does there exist a function h such that $\varphi = dh$?" In order to answer this question, we compare

$$dh = \frac{\partial h}{\partial u}\,du + \frac{\partial h}{\partial v}\,dv \tag{2.5.10}$$

with

$$\varphi = f\,du + g\,dv. \tag{2.5.11}$$

The function h we are seeking must satisfy

$$f = \frac{\partial h}{\partial u}, \qquad g = \frac{\partial h}{\partial v}. \tag{2.5.12}$$

From the first equation of (2.5.12) we must have

$$h(u, v) = \int_a^u f(x, v)\,dx + \alpha(v). \tag{2.5.13}$$

On the other hand, for the second equation of (2.5.12) to hold we should have

$$g(u, v) = \int_a^u \frac{\partial f(x, v)}{\partial v}\,dx + \frac{d\alpha(v)}{dv}. \tag{2.5.14}$$

From (2.5.3), $d\varphi = 0$ amounts to

$$\frac{\partial f}{\partial v} = \frac{\partial g}{\partial u}, \tag{2.5.15}$$

so (2.5.14) becomes

$$\begin{aligned} g(u, v) &= \int_a^u \frac{\partial g(x, v)}{\partial x}\,dx + \frac{d\alpha(v)}{dv} \\ &= g(u, v) - g(a, v) + \frac{d\alpha(v)}{dv}. \end{aligned} \tag{2.5.16}$$

Thus, if we define

$$\alpha(v) = \int_c^v g(a, y)\,dy, \tag{2.5.17}$$

the function h defined by (2.5.13) satisfies $\varphi = dh$. However, in the above argument, in order to define $h(u, v)$, both $f(x, v)$ and $g(a, y)$ had to be integrable for $a \leq x \leq u$ and $c \leq y \leq v$. Thus

Theorem 2.5.1 *If a 1-form*

$$\varphi = f(u, v)\, du + g(u, v)\, dv$$

is continuously differentiable in a region $a \leq u \leq b$, $c \leq v \leq d$ and if $d\varphi = 0$, then there exists a function h defined in that region such that $\varphi = dh$.

This is one case of **Poincaré's lemma**. The assumption on the domain of φ is very important. For example, the 1-form

$$\varphi = \frac{-v\, du + u\, dv}{u^2 + v^2} \tag{2.5.18}$$

is not defined at the origin $(0, 0)$, but $d\varphi = 0$ at other points. Now, assume that there is a function h defined in the region $0 < u^2 + v^2 < a^2$ around the origin such that $\varphi = dh$. We shall show that this assumption leads to a contradiction. Take a small ε $(0 < \varepsilon < a)$ and integrate φ along the circle $C : u^2 + v^2 = \varepsilon^2$ of radius ε. Since the starting point $(\varepsilon, 0)$ of C is equal to the ending point and $\varphi = dh$, we have

$$\int_C \varphi = \Big[h\Big]_{(\varepsilon,0)}^{(\varepsilon,0)} = 0. \tag{2.5.19}$$

On the other hand, using the polar coordinates r, θ $(u = r\cos\theta, v = r\sin\theta)$, we rewrite (2.5.18):

$$\varphi = d\theta. \tag{2.5.20}$$

When a point goes around C once, θ changes from 0 to 2π. Thus we have

$$\int_C d\theta = 2\pi, \tag{2.5.21}$$

which contradicts (2.5.19). Here it seems we have $\varphi = d\theta$, with a function θ defined on the plane minus the origin. However, actually, θ is not uniquely defined. For example, on the positive real axis θ can be 0, 2π, 4π, etc.. With the condition $0 \leq \theta < 2\pi$ the function θ is uniquely defined, but then it would be discontinuous along the positive real axis.

A 2-form $\psi = f\; du \wedge dv$ always satisfies $d\psi = 0$. Poincaré's lemma holds also in this case. Thus,

Theorem 2.5.2 *If a 2-form*

$$\psi = f\; du \wedge dv$$

is continuously differentiable in a region $a \leq u \leq b$, $c \leq v \leq d$, then there exists a 1-form φ defined in that region such that $\psi = d\varphi$.

The proof for 2-forms is much easier than that of Theorem 2.5.1. We just have to define φ as follows:

$$\varphi = g\; dv, \quad \text{where } g(u, v) = \int f(u, v) du.$$

Next, we shall see how differential forms under the variable change:

$$u = \varphi(s,t), \quad v = \psi(s,t). \tag{2.5.22}$$

By (2.5.22) a 1-form

$$a\,du + b\,dv \tag{2.5.23}$$

transforms to

$$a\left(\frac{\partial u}{\partial s}ds + \frac{\partial u}{\partial t}dt\right) + b\left(\frac{\partial v}{\partial s}ds + \frac{\partial v}{\partial t}dt\right). \tag{2.5.24}$$

(More exactly, a in (2.5.23) is a function $a(u,v)$ of (u,v), and a in (2.5.24) is a function $a(\varphi(s,t),\psi(s,t))$ of (s,t).) Under the same variable change a 2-form

$$a\,du \wedge dv \tag{2.5.25}$$

transforms to

$$a\left(\frac{\partial u}{\partial s}ds + \frac{\partial u}{\partial t}dt\right) \wedge \left(\frac{\partial v}{\partial s}ds + \frac{\partial v}{\partial t}dt\right). \tag{2.5.26}$$

Simplify this using (2.5.1). Then we can rewrite (2.5.26) as follows:

$$a\left(\frac{\partial u}{\partial s}\frac{\partial v}{\partial t} - \frac{\partial u}{\partial t}\frac{\partial v}{\partial s}\right) ds \wedge dt = a\begin{vmatrix} \dfrac{\partial u}{\partial s} & \dfrac{\partial u}{\partial t} \\ \dfrac{\partial v}{\partial s} & \dfrac{\partial v}{\partial t} \end{vmatrix} ds \wedge dt. \tag{2.5.27}$$

This agrees with the transformation formula for the double integral:

$$\iint a(u,v)\,dudv = \iint a(u(s,t),v(s,t))\begin{vmatrix} \dfrac{\partial u}{\partial s} & \dfrac{\partial u}{\partial t} \\ \dfrac{\partial v}{\partial s} & \dfrac{\partial v}{\partial t} \end{vmatrix} dsdt, \tag{2.5.28}$$

which we learned in calculus. This shows that $dudv$ in the double integral should be understood as $du \wedge dv$. Readers should appreciate how simple it was to calculate (2.5.27) compared with the usual proof of (2.5.28) in calculus. Historically, exterior differential forms started in connection with multiple integrals.

Problem 2.5.1 When we change the coordinates (u,v) to the polar coordinates (r,θ) by $u = r\cos\theta$, $v = r\sin\theta$, prove that

$$du \wedge dv = r\,dr \wedge d\theta.$$

(This agrees with a familiar formula for double integrals in calculus.)

2.6 Use of Exterior Differential Forms

Using differential forms introduced in the preceding section, we review the theory of surfaces we studied in Sections 2.2 and 2.4. As in (2.2.1), we write a surface as follows:

$$p(u, v) = (x(u, v), y(u, v), z(u, v)). \tag{2.6.1}$$

As in (2.4.10), we write $\boldsymbol{dp} = (dx, dy, dz)$ as

$$\boldsymbol{dp} = \theta^1 \boldsymbol{e}_1 + \theta^2 \boldsymbol{e}_2, \tag{2.6.2}$$

where $(\boldsymbol{e}_1, \boldsymbol{e}_2)$ ia an orthonormal frame, and θ^1 and θ^2 are 1-forms given by (2.4.9). Since the exterior differentiation d defined in the preceding section satisfies $\boldsymbol{ddp} = (ddx, ddy, ddz) = \boldsymbol{0}$ from (2.6.2) and (2.4.12) we obtain

$$\begin{aligned} \boldsymbol{0} &= d\theta^1 \boldsymbol{e}_1 - \theta^1 \wedge \sum_{j=1}^{3} {\omega_1}^j \boldsymbol{e}_j + d\theta^2 \boldsymbol{e}_2 - \theta^2 \wedge \sum_{j=1}^{3} {\omega_2}^j \boldsymbol{e}_j \\ &= \left(d\theta^1 - \sum_{j=1}^{2} \theta^j \wedge {\omega_j}^1 \right) \boldsymbol{e}_1 + \left(d\theta^2 - \sum_{j=1}^{2} \theta^j \wedge {\omega_j}^2 \right) \boldsymbol{e}_2 - \sum_{j=1}^{2} \theta^j \wedge {\omega_j}^3 \boldsymbol{e}_3. \end{aligned} \tag{2.6.3}$$

(In the calculation above, note that from (2.5.7) we get the minus sign in $d(\theta^1 \boldsymbol{e}_1) = d\theta^1 \boldsymbol{e}_1 - \theta^1 \wedge \boldsymbol{de}_1$. Note also that just as in the case of $\boldsymbol{dp}$, $\boldsymbol{de}_1$ denotes the exterior derivative of the components of $\boldsymbol{e}_1$.) Since $\boldsymbol{e}_1$, $\boldsymbol{e}_2$, and $\boldsymbol{e}_3$ are linearly independent, from (2.6.3) we get

$$d\theta^i = \sum_{j=1}^{2} \theta^i \wedge {\omega_j}^i \quad (i = 1, 2), \tag{2.6.4}$$

$$0 = \sum_{j=1}^{2} \theta^i \wedge {\omega_j}^3. \tag{2.6.5}$$

We call (2.6.4) the **first structure equation** of the surface. As in (2.4.16), write

$${\omega_j}^3 = \sum_{k=1}^{2} b_{jk} \theta^k \quad (j = 1, 2), \tag{2.6.6}$$

and substituting this into (2.6.5) we get

$$0 = \sum_{j,k=1}^{2} b_{jk} \theta^j \wedge \theta^k. \tag{2.6.7}$$

Since $\theta^1 \wedge \theta^1 = \theta^2 \wedge \theta^2 = 0$ and $\theta^1 \wedge \theta^2 = -\theta^2 \wedge \theta^1$ by (2.5.1), (2.6.7) can be written as

$$0 = b_{12} \theta^1 \wedge \theta^2 + b_{21} \theta^2 \wedge \theta^1 = (b_{12} - b_{21}) \theta^1 \wedge \theta^2. \tag{2.6.8}$$

Since θ^1 and θ^2 are linearly independent so that, $\theta^1 \wedge \theta^2 \neq 0$, we have

$$b_{12} = b_{21}, \tag{2.6.9}$$

verifying once again the symmetricity of the matrix B proved in (2.4.18). We saw in (2.4.14) that the matrix with entries ${\omega_j}^i$ is skew-symmetric, i.e., ${\omega_1}^1 = {\omega_2}^2 = 0$ and ${\omega_2}^1 = -{\omega_1}^2$. Thus, the first structure equation (2.6.4) is nothing but

$$
\begin{aligned}
d\theta^1 &= \theta^2 \wedge \omega_2{}^1, \\
d\theta^2 &= \theta^1 \wedge \omega_1{}^2.
\end{aligned}
\tag{2.6.10}
$$

Next, in the same way as we obtained (2.6.3) from $dd\boldsymbol{p} = \boldsymbol{0}$, we derive from $dd\boldsymbol{e}_i = \boldsymbol{0}$ the following:

$$
\begin{aligned}
\boldsymbol{0} = d(d\boldsymbol{e}_i) &= d\left(\sum_{j=1}^{3} \omega_i{}^j \boldsymbol{e}_j\right) \\
&= \sum_{j=1}^{3} d\omega_i{}^j \boldsymbol{e}_j - \sum_{j=1}^{3} \omega_i{}^j \wedge d\boldsymbol{e}_j \\
&= \sum_{k=1}^{3} \left(d\omega_i{}^k - \sum_{j=1}^{3} \omega_i{}^j \wedge \omega_j{}^k\right) \boldsymbol{e}_k.
\end{aligned}
\tag{2.6.11}
$$

In particular, for $i = 1, 2$, looking at the coefficients of $\boldsymbol{e}_k$ $(k = 1, 2)$, we have

$$
\begin{aligned}
d\omega_i{}^k &= \sum_{j=1}^{3} \omega_i{}^j \wedge \omega_j{}^k \\
&= \sum_{j=1}^{2} \omega_i{}^j \wedge \omega_j{}^k + \omega_i{}^3 \wedge \omega_3{}^k \\
&= \sum_{j=1}^{2} \omega_i{}^j \wedge \omega_j{}^k + \omega_k{}^3 \wedge \omega_i{}^3 \\
&= \sum_{j=1}^{2} \omega_i{}^j \wedge \omega_j{}^k + \sum_{h,j=1}^{2} b_{kh} b_{ij} \theta^h \wedge \theta^j \\
&= \sum_{j=1}^{2} \omega_i{}^j \wedge \omega_j{}^k + \frac{1}{2} \sum_{h,j=1}^{2} (b_{kh} b_{ij} - b_{kj} b_{ih}) \theta^h \wedge \theta^j.
\end{aligned}
\tag{2.6.12}
$$

Since $[\omega_i{}^k]$ is skew-symmetric, we need to consider the equation above only for $k = 1, i = 2$. In that case,

$$
\begin{aligned}
d\omega_2{}^1 &= \sum_{j=1}^{2} \omega_2{}^j \wedge \omega_j{}^1 + \frac{1}{2} \sum_{h,j=1}^{2} (b_{1h} b_{2j} - b_{1j} b_{2h}) \theta^h \wedge \theta^j \\
&= (b_{11} b_{22} - b_{12} b_{21}) \theta^1 \wedge \theta^2.
\end{aligned}
\tag{2.6.13}
$$

Using (2.4.24) we can rewrite (2.6.13) as

$$
d\omega_2{}^1 = K\, \theta^1 \wedge \theta^2, \quad K = b_{11} b_{22} - b_{12} b_{21}. \tag{2.6.14}
$$

We call (2.6.14) the **second structure equation** of the surface. This will reappear as Gauss' theorem in the next chapter.

We obtained (2.6.13) from the coefficients of $\boldsymbol{e}_1$, $\boldsymbol{e}_2$ in (2.6.11). Now we look at the coefficient of $\boldsymbol{e}_3$. From (2.6.11) we get

$$d\omega_i{}^3 - \sum_{j=1}^{3} \omega_i{}^j \wedge \omega_j{}^3 = 0 \quad (i = 1, 2). \tag{2.6.15}$$

Since $\omega_3{}^3 = 0$, the last term vanishes, and we get

$$d\omega_i{}^3 - \sum_{j=1}^{2} \omega_i{}^j \wedge \omega_j{}^3 = 0. \tag{2.6.16}$$

Substituting (2.6.6) into this we have

$$d\left(\sum_{j=1}^{2} b_{ij}\theta^j\right) - \sum_{j=1}^{2} \omega_i{}^j \wedge \left(\sum_{k=1}^{2} b_{jk}\theta^k\right) = 0. \tag{2.6.17}$$

Using (2.6.4) we can calculate the first term as follows:

$$\begin{aligned} d\left(\sum_{j=1}^{2} b_{ij}\theta^j\right) &= \sum_{j=1}^{2} db_{ij} \wedge \theta^j - \sum_{j,k=1}^{2} b_{ij}\omega_k{}^j \wedge \theta^k \\ &= \sum_{k=1}^{2}\left(db_{ik} - \sum_{j=1}^{2} b_{ij}\omega_k{}^j\right) \wedge \theta^k. \end{aligned} \tag{2.6.18}$$

Substituting this into (2.6.17) we get

$$\sum_{k=1}^{2}\left(db_{ik} - \sum_{j=1}^{2} b_{ij}\omega_k{}^j - \sum_{j=1}^{2} b_{jk}\omega_i{}^j\right) \wedge \theta^k = 0. \tag{2.6.19}$$

Since the terms inside ( ) are 1-forms, they are linear combinations of θ^1 and θ^2. Thus,

$$db_{ik} - \sum_{j=1}^{2} b_{ij}\omega_k{}^j - \sum_{j=1}^{2} b_{jk}\omega_i{}^j = \sum_{l=1}^{2} b_{ik,l}\theta^l. \tag{2.6.20}$$

We define the coefficients $b_{ik,l}$ for the linear combinations by the equations above. We write $b_{ik,l}$ instead of b_{ikl} following the tradition of tensor analysis. Now we interchange i and k in (2.6.20) and compare the resulting equation with (2.6.20) itself. Then from $b_{ik} = b_{ki}$, we get $b_{ik,l} = b_{ki,l}$. Substituting the right-hand side of (2.6.20) into (2.6.19) we get

$$\sum_{k,l=1}^{2} b_{ik,l}\theta^l \wedge \theta^k = 0 \quad (i = 1, 2). \tag{2.6.21}$$

Since $\theta^1 \wedge \theta^1 = \theta^2 \wedge \theta^2 = 0$ and $\theta^1 \wedge \theta^2 = -\theta^2 \wedge \theta^1$, from (2.6.21) we get $b_{i2,1} = b_{i1,2}$. Hence,

$$b_{ik,l} = b_{il,k}. \tag{2.6.22}$$

We call this **Mainardi-Codazzi's equation.** Since $b_{ik,l}$ is symmetric also in i and k, it is symmetric in all three indices i, j, k. We call (2.6.14) together with (2.6.22) the **fundamental equations of the theory of surfaces.**

In Section 1.4 of Chapter 1, we proved that given functions $\kappa > 0$ and τ, there exists a space curve with curvature κ and torsion τ and that such a curve is unique. Now we consider a similar question for surfaces. Consider a region D in the (u, v)-plane. For simplicity, assume that it is a rectangular region $|u| < a$, $|v| < b$. Let θ^1 and θ^2 be 1-forms which are everywhere linearly independent in D. Given

$$\theta^1\theta^1 + \theta^2\theta^2, \tag{2.6.23}$$

$$\sum_{i,j=1}^{2} b_{ij}\theta^i\theta^j \qquad \text{with } b_{ij} = b_{ji}, \tag{2.6.24}$$

does there exist a surface $\boldsymbol{p}(u, v)$ which has (2.6.23) and (2.6.24) as its first and second fundamental forms? In this section we explain the answer to this question without giving a proof. Since we will not use what follows in the rest of the book, readers do not have to worry about understanding it completely.

First, given (2.6.23), there exists on D, 1-forms ${\omega_j}^i$ satisfying

$${\omega_j}^i + {\omega_i}^j = 0 \quad (i, j = 1, 2), \tag{2.6.25}$$

$$d\theta^i = -\sum_{j=1}^{2} {\omega_j}^i \wedge \theta^j \quad (i = 1, 2). \tag{2.6.26}$$

In (2.4.14) and (2.6.4) we proved that such ${\omega_j}^i$'s exist for a surface in the space. Actually, from θ^1, θ^2 alone, we can obtain a unique set of 1-forms ${\omega_j}^i$ satisfying (2.6.25) and (2.6.26). We will prove this fact in the next section. The set of 1-forms ${\omega_j}^i$ being determined by θ^1, θ^2, we define a function K by the following equation:

$$d{\omega_2}^1 = K\theta^1 \wedge \theta^2. \tag{2.6.27}$$

If there exists a surface $\boldsymbol{p}(u, v)$, such that (2.6.23) and (2.6.24) are its first and second fundamental forms, the second equation of (2.6.14)

$$K = b_{11}b_{22} - b_{12}b_{21} \tag{2.6.28}$$

must hold. Hence, (2.6.28) is a necessary condition. Next, if we define $b_{ik,l}$ as in (2.6.20), we have proved in (2.6.22) that

$$b_{i2,1} - b_{i1,2} = 0 \tag{2.6.29}$$

must hold. Hence this is another necessary condition. The fundamental theorem of surfaces states:

Conversely, if (2.6.28) and (2.6.29) hold, there exists a surface in the space such that (2.6.23) and (2.6.24) are its first and second fundamental forms, and it is unique up to a congruence.

This is a fairly easy problem on partial differential equation, and (2.6.28) and (2.6.29) are called its integrability conditions. In case of curves, we needed only ordinary differential equations. However for surfaces, because there are two variables u and v, we need partial differential equations. With a little more of the general theory of differential forms, the proof would be easy. However, we shall not go any further in this book.

Problem 2.6.1 As in Example 2.1.1, let

$$\begin{aligned} \boldsymbol{e}_1 &= (-\sin u \cos v, -\sin u \sin v, \cos u), \\ \boldsymbol{e}_2 &= (-\sin v, \cos v, 0) \end{aligned}$$

be a parametric representation of a sphere of radius a. Express $\boldsymbol{e}_3$, θ^1, θ^2 and $\omega_j{}^i$ $(1 \leq i, j \leq 3)$ in terms of u and v. Then calculate $d\omega_2{}^1 = K\theta^1 \wedge \theta^2$ to find K. Also, check that

$$\begin{bmatrix} b_{11} & b_{12} \\ b_{21} & b_{22} \end{bmatrix} = \begin{bmatrix} \dfrac{1}{a} & 0 \\ 0 & \dfrac{1}{a} \end{bmatrix}.$$

(Here, $\boldsymbol{e}_1$ is the unit tangent vector of the curve with parameter u and fixed v, and $\boldsymbol{e}_2$ is the unit tangent vector of the curve with parameter v and fixed u.)

Problem 2.6.2 Define parameters (u, v) for a torus as in Example 2.1.7, Let $\boldsymbol{e}_1$ be the unit tangent vector of the curve with parameter u and fixed v, and $\boldsymbol{e}_2$ the unit tangent vector of the curve with parameter v and fixed u. Calculate $\boldsymbol{e}_3$, θ^1, θ^2, $\omega_j{}^i$, K, and b_{ij} as in Problem 2.6.1.

Problem 2.6.3 Use the notation in Problem 2.4.1, and write L_{11}, $L_{12} = L_{21}$, L_{22} for L, M, N. Thus the second fundamental form becomes

$$\sum_{i,j=1}^{2} L_{ij} du^i du^j.$$

Next, define

$$L_{ij,k} = \frac{\partial L_{ij}}{\partial u^k} - \sum_{l=1}^{2} \Gamma^l_{ik} L_{lj} - \sum_{l=1}^{2} \Gamma^l_{jk} L_{il}.$$

Then, prove that

$$L_{ij,k} - L_{ik,j} = 0.$$

Hint: This is nothing but Mainardi-Codazzi's equation (2.6.22) rewritten in the notation of Section 2.2, and can be obtained from the normal component of $dd\boldsymbol{p}_i = \boldsymbol{0}$. The point is to use (2.2.25) when calculating $dd\boldsymbol{p}_i$.

Chapter 3
Geometry of Surfaces

3.1 Riemannian Metrics on a Surface

In this chapter we will study properties of a surface that depend only on the first fundamental form. Take, for example, the cylindrical surface

$$x = x(u), \quad y = y(u), \quad z = v. \tag{3.1.1}$$

Assume that as in Example 2.3.1 of Chapter 2, the parameter u is set so that

$$\left(\frac{dx}{du}\right)^2 + \left(\frac{dy}{du}\right)^2 = 1. \tag{3.1.2}$$

Then, as we have already checked, the first fundamental form is given by

$$\mathbf{I} = (du)^2 + (dv)^2. \tag{3.1.3}$$

Thus as long as (3.1.2) holds, the first fundamental form of any cylindrical surface (3.1.1) is equal to (3.1.3), which is the first fundamental form of a plane. As we calculated in Example 2.3.1 of Chapter 2, the mean curvature H varies with (3.1.1), so it is not determined by the first fundamental form $\mathbf{I}$ alone.

On the other hand, the Gaussian curvature K is 0 for all cylindrical surfaces, and there arises the possibility that it would depend only on the first fundamental form $\mathbf{I}$. Indeed, this is the case as we will see later. The purpose of this chapter is to examine those quantities and properties that can be determined by the first fundamental form alone. In other words, we shall forget that the surface is a part of the 3-dimensional Euclidean space, and develop the theory using only the first fundamental form.

Thus, in a domain D of the (u, v)-plane, consider

$$E\,dudu + 2F\,dudv + G\,dvdv, \tag{3.1.4}$$

which is to play the role of the first fundamental form. Since we will use (3.1.4) to measure the length of a curve $(u(t), v(t))$ in D as in (2.2.13),

---

The original version of the chapter was revised: Belated corrections have been incorporated. The correction to the chapter is available at https://doi.org/10.1007/978-981-15-1739-6_6

S. Kobayashi, *Differential Geometry of Curves and Surfaces*,
Springer Undergraduate Mathematics Series,
https://doi.org/10.1007/978-981-15-1739-6_3

$$E\left(\frac{du}{dt}\right)^2 + 2F\frac{du}{dt}\frac{dv}{dt} + G\left(\frac{dv}{dt}\right)^2 \tag{3.1.5}$$

has to be always positive unless

$$\frac{du}{dt} = \frac{dv}{dt} = 0.$$

Hence, the quadratic form (3.1.4) has to be positive definite. This condition can also be given by

$$EG - F^2 > 0, \quad E > 0. \tag{3.1.6}$$

When (3.1.4) is a positive definite form, we call it a **Riemannian metric** on D. This is in honor of Riemann, who proposed studies of geometry based only on the first fundamental form at the inaugural address at Göttingen University on June 10, 1854 at the age of 27. (Riemann spoke not only about the 2-dimensional case, but also about the general n-dimensional case.)

In (3.1.5), let the length of the curve be used as the parameter. Then, we usually use s instead of t, and (2.2.13) becomes

$$\int_\alpha^\beta \sqrt{E\left(\frac{du}{ds}\right)^2 + 2F\frac{du}{ds}\frac{dv}{ds} + G\left(\frac{dv}{ds}\right)^2}\, ds. \tag{3.1.7}$$

Since s is the parameter which represents length, we get

$$E\left(\frac{du}{ds}\right)^2 + 2F\frac{du}{ds}\frac{dv}{ds} + G\left(\frac{dv}{ds}\right)^2 = 1. \tag{3.1.8}$$

Formally, (3.1.8) may be expressed as

$$ds^2 = E\, dudu + 2F\, dudv + G\, dvdv, \tag{3.1.9}$$

so we usually write the Riemannian metric (3.1.4) as ds^2.

Of course, the the first fundamental form of a surface discussed in Section 2.2 of Chapter 2 is an important example of Riemannian metrics, but here we will give two examples of Riemannian metrics defined directly in a domain D, which are important for application.

Example 3.1.1 Let D be the interior of a unit circle defined by $u^2 + v^2 < 1$ and let

$$ds^2 = 4\frac{(du)^2 + (dv)^2}{\{1 - (u^2 + v^2)\}^2}. \tag{3.1.10}$$

We call this the **Poincaré metric**. If we use the complex coordinate $w = u + iv$, we can write the above metric as

$$ds^2 = \frac{4dwd\bar{w}}{(1 - |w|^2)^2}, \tag{3.1.11}$$

where $dw = du + idv$, $d\bar{w} = du - idv$. ♦

Example 3.1.2 Let U be the upper-half plane defined by $\{(x, y); y > 0\}$. In terms of the complex coordinate $z = x + iy$, U is the set of complex numbers with positive imaginary part. On U, consider a Riemannian metric

$$ds^2 = \frac{(dx)^2 + (dy)^2}{y^2} = \frac{dz\,d\bar{z}}{y^2}. \tag{3.1.12}$$

We call this also the **Poincaré metric.** ♦

Let us explain that the above two examples are essentially the same, although they look different. As some readers may know from complex function theory, the transformation

$$z = i\,\frac{1-w}{1+w} \tag{3.1.13}$$

gives a one-to-one correspondence between $D = \{w\,;\,|w| < 1\}$ and $U = \{z = x + iy\,;\, y > 0\}$. In fact, since

$$2yi = z - \bar{z} = i\,\frac{1-w}{1+w} + i\,\frac{1-\bar{w}}{1+\bar{w}} = \frac{2(1-w\bar{w})}{(1+w)(1+\bar{w})}\,i, \tag{3.1.14}$$

we have $y > 0$ if and only if $|w|^2 < 1$. Also, solving (3.1.13) for w, we get the inverse map of (3.1.13):

$$w = \frac{i-z}{i+z}. \tag{3.1.15}$$

Thus, we see that (3.1.13) and (3.1.15) give a one-to-one complex analytic correspondence between D and U. From (3.1.15) we get

$$dw = \frac{-2i\,dz}{(i+z)^2}, \quad d\bar{w} = \frac{2i\,d\bar{z}}{(-i+\bar{z})^2}, \quad \frac{1}{(1-|w|^2)^2} = \frac{(i+z)^2(-i+\bar{z})^2}{(4y)^2}, \tag{3.1.16}$$

and by substituting these into (3.1.11) we get

$$\frac{4dw\,d\bar{w}}{(1-|w|^2)^2} = \frac{dz\,d\bar{z}}{y^2}. \tag{3.1.17}$$

This shows that under the correspondence between D and U given by (3.1.13) or (3.1.15), the Riemannian metric (3.1.11) on D corresponds to the Riemannian metric (3.1.12) on U.

In general, given a domain D in the (u, v)-plane with a Riemannian metric

$$E\,dudu + 2F\,dudv + G\,dvdv, \tag{3.1.18}$$

and a domain D' in the (x, y)-plane with a Riemannian metric

$$E'\,dxdx + 2F'\,dxdy + G'\,dydy, \tag{3.1.19}$$

if there is a one-to-one correspondence between D and D' by differentiable functions

$$u = u(x, y), \qquad v = v(x, y), \tag{3.1.20}$$

which, substituted into (3.1.18), yields (3.1.19), then we call the correspondence an **isometry**. If (3.1.18) and (3.1.19) correspond to each other under an isometry, then they are regarded essentially the same.

An important problem is whether a Riemannian metric which is defined on a domain in the plane, such as (3.1.11) or (3.1.12), can be obtained as the first fundamental form of a surface in the space. This is called the **isometric imbedding** problem.

Problem 3.1.1 Compare the first and second fundamental forms, H, K, κ_1, and κ_2 of a catenoid

$$\boldsymbol{p}(u, v) = \left(\sqrt{u^2 + a^2}\,\cos v,\ \sqrt{u^2 + a^2}\,\sin v,\ a \sinh^{-1}\frac{u}{a}\right)$$

with those of a right helicoid

$$\boldsymbol{q}(u, v) = (u \cos v,\ u \sin v,\ av).$$

3.2 Structure Equations of a Surface

On a domain D in the (u, v)-plane, consider a Riemannian metric (3.1.9). As in (2.4.11) of Chapter 2, we shall first prove that it can be written as

$$ds^2 = \theta^1\theta^1 + \theta^2\theta^2, \tag{3.2.1}$$

using 1-forms θ^1 and θ^2 which are linearly independent on D. First, let

$$\theta^1 = a_1{}^1\, du + a_2{}^1\, dv, \qquad \theta^2 = a_1{}^2\, du + a_2{}^2\, dv. \tag{3.2.2}$$

Then, (3.2.1) is equivalent to

$$\begin{aligned} E &= (a_1{}^1)^2 + (a_1{}^2)^2, \\ F &= a_1{}^1\, a_2{}^1 + a_1{}^2\, a_2{}^2, \\ G &= (a_2{}^1)^2 + (a_2{}^2)^2. \end{aligned} \tag{3.2.3}$$

Thus it suffices to set

$$\begin{aligned} a_1{}^1 &= \sqrt{E}, & a_1{}^2 &= 0, \\ a_2{}^1 &= \frac{F}{\sqrt{E}}, & a_2{}^2 &= \sqrt{\frac{EG - F^2}{E}}. \end{aligned} \tag{3.2.4}$$

In the sequel, as long as θ^1, θ^2 satisfy (3.2.1) they need not to be given by (3.2.4).

Next, we search for $\omega_2{}^1 (= -\omega_1{}^2)$ such that the first structure equation (2.6.4), i.e.,

$$d\theta^1 = \theta^2 \wedge \omega_2{}^1, \quad d\theta^2 = \theta^1 \wedge \omega_1{}^2, \quad \text{where } \omega_1{}^2 = -\omega_2{}^1, \tag{3.2.5}$$

holds. Since $\omega_2{}^1$ we are seeking is a 1-form, let

$$\omega_2{}^1 = b_1\,\theta^1 + b_2\,\theta^2. \tag{3.2.6}$$

Then (3.2.5) becomes

$$d\theta^1 = -b_1\,\theta^1 \wedge \theta^2, \quad d\theta^2 = -b_2\,\theta^1 \wedge \theta^2. \tag{3.2.7}$$

By (3.2.7), b_1 and b_2 are uniquely determined. Thus, we get a unique skew-symmetric matrix of 1-forms,

$$\omega = \begin{bmatrix} 0 & \omega_2{}^1 \\ \omega_1{}^2 & 0 \end{bmatrix}, \qquad \omega_1{}^2 = -\omega_2{}^1 \tag{3.2.8}$$

for which (3.2.5) holds. We call this matrix valued 1-form the **connection form**.

As in (2.6.14) of Chapter 2, we define K by the second structure equation

$$d\omega_2{}^1 = K\,\theta^1 \wedge \theta^2, \tag{3.2.9}$$

and call it the **Gaussian curvature**.

In the calculation above, we chose θ^1, θ^2 in such a way that (3.2.1) holds. Now, we shall see what happens when we choose other $\bar\theta^1, \bar\theta^2$ for which (3.2.1) still holds. We set

$$\begin{aligned} \bar\theta^1 &= s_1{}^1\,\theta^1 + s_2{}^1\,\theta^2, \\ \bar\theta^2 &= s_1{}^2\theta^1 + s_2{}^2\,\theta^2. \end{aligned} \tag{3.2.10}$$

In the matrix notation,

$$\begin{bmatrix} \bar\theta^1 \\ \bar\theta^2 \end{bmatrix} = \begin{bmatrix} s_1{}^1 & s_2{}^1 \\ s_1{}^2 & s_2{}^2 \end{bmatrix} \begin{bmatrix} \theta^1 \\ \theta^2 \end{bmatrix}, \tag{3.2.11}$$

or simply, $\bar\theta = S\theta$. Transposing (3.2.11) we have

$$(\bar\theta^1, \bar\theta^2) = (\theta^1, \theta^2) \begin{bmatrix} s_1{}^1 & s_1{}^2 \\ s_2{}^1 & s_2{}^2 \end{bmatrix}, \tag{3.2.12}$$

or simply, ${}^t\bar\theta = {}^t\theta\,{}^tS$. Since

$$\bar\theta^1\bar\theta^1 + \bar\theta^2\bar\theta^2 = (\bar\theta^1, \bar\theta^2) \begin{bmatrix} \bar\theta^1 \\ \bar\theta^2 \end{bmatrix} = {}^t\theta\,{}^tSS\theta \tag{3.2.13}$$

and

$$\bar\theta^1\bar\theta^1 + \bar\theta^2\bar\theta^2 = \theta^1\theta^1 + \theta^2\theta^2, \quad \text{i.e., } {}^t\bar\theta\,\bar\theta = {}^t\theta\,\theta, \tag{3.2.14}$$

we have

$${}^tSS = I, \quad \text{where } I = \begin{bmatrix} 1 & 0 \\ 0 & 1 \end{bmatrix}. \tag{3.2.15}$$

This shows that S is an orthogonal matrix. Calculating the exterior derivative of (3.2.11), we get

$$d\bar\theta = dS \wedge \theta + S\,d\theta. \tag{3.2.16}$$

On the other hand, from

$$\begin{bmatrix} d\theta^1 \\ d\theta^2 \end{bmatrix} = -\begin{bmatrix} 0 & \omega_2{}^1 \\ \omega_1{}^2 & 0 \end{bmatrix} \wedge \begin{bmatrix} \theta^1 \\ \theta^2 \end{bmatrix}, \tag{3.2.17}$$

or in the matrix notation,

$$d\theta = -\omega \wedge \theta, \qquad \omega = \begin{bmatrix} 0 & {\omega_2}^1 \\ {\omega_1}^2 & 0 \end{bmatrix}, \tag{3.2.18}$$

substituting (3.2.11) and (3.2.18) into (3.2.16) yields

$$\begin{aligned} d\bar{\theta} &= dS \cdot S^{-1} \wedge \bar{\theta} - S\omega S^{-1} \wedge \bar{\theta} \\ &= -(-dS \cdot S^{-1} + S\omega S^{-1}) \wedge \bar{\theta}. \end{aligned} \tag{3.2.19}$$

If we set

$$\bar{\omega} = -dS \cdot S^{-1} + S\omega S^{-1}, \tag{3.2.20}$$

then (3.2.19) can be written as follows:

$$d\bar{\theta} = -\bar{\omega} \wedge \bar{\theta}, \tag{3.2.21}$$

which is the first structure equation (3.2.18) when $(\bar{\theta}^1, \bar{\theta}^2)$ is used in place of (θ^1, θ^2). However, we need to check that $\bar{\omega}$ is skew-symmetric. From

$$ {}^tS = S^{-1}, \quad {}^t\omega = -\omega,$$

$$0 = d(SS^{-1}) = dS \cdot S^{-1} + S \cdot d(S^{-1}),$$

we obtain

$$\begin{aligned} {}^t\bar{\omega} &= -{}^tS^{-1} \cdot d\,{}^tS + {}^tS^{-1} \cdot {}^t\omega \cdot {}^tS \\ &= -S \cdot d(S^{-1}) - S \cdot \omega \cdot S^{-1} \\ &= dS \cdot S^{-1} - S \cdot \omega \cdot S^{-1} \\ &= -\bar{\omega}, \end{aligned} \tag{3.2.22}$$

showing that $\bar{\omega}$ is skew-symmetric.

Thus, we have shown that the connection forms $\omega = [\,{\omega_j}^i\,]$ and $\bar{\omega} = [\,{\bar{\omega}_j}^i\,]$ corresponding to (θ^1, θ^2) and $(\bar{\theta}^1, \bar{\theta}^2)$, respectively, are related by (3.2.20). Next, in order to see how the change from (θ^1, θ^2) to $(\bar{\theta}^1, \bar{\theta}^2)$ affects the Gaussian curvature, we compare $d\omega$ with $d\bar{\omega}$. We multiply (3.2.20) by S from the right and exterior differentiate. Then

$$d(\bar{\omega}S) = d(-dS + S\omega). \tag{3.2.23}$$

The left-hand side is equal to

$$\begin{aligned} d(\bar{\omega}S) &= d\bar{\omega} \cdot S - \bar{\omega} \wedge dS \\ &= d\bar{\omega} \cdot S + dS \cdot S^{-1} \wedge dS - S \cdot \omega \cdot S^{-1} \wedge dS. \end{aligned} \tag{3.2.24}$$

On the other hand, the right-hand side of (3.2.23) is equal to

$$d(-dS + S \cdot \omega) = dS \wedge \omega + S \cdot d\omega. \tag{3.2.25}$$

Since the right-hand sides of (3.2.24) and (3.2.25) are equal, multiplying them by S^{-1} from the right yields

$$d\bar{\omega} + dS \cdot S^{-1} \wedge dS \cdot S^{-1} - S \cdot \omega \cdot S^{-1} \wedge dS \cdot S^{-1} = dS \wedge \omega \cdot S^{-1} + S \cdot d\omega \cdot S^{-1}, \quad (3.2.26)$$

which may be rewritten as

$$d\bar{\omega} = S \cdot d\omega \cdot S^{-1} - dS \cdot S^{-1} \wedge dS \cdot S^{-1} + S \cdot \omega \cdot S^{-1} \wedge dS \cdot S^{-1} + dS \cdot S^{-1} \wedge S \cdot \omega \cdot S^{-1}. \quad (3.2.27)$$

Since S is an orthogonal matrix,

$${}^t(dS \cdot S^{-1}) = {}^tS^{-1} \cdot d\,{}^tS = S \cdot d(S^{-1}) = d(SS^{-1}) - dS \cdot S^{-1} = -dS \cdot S^{-1}, \quad (3.2.28)$$

which shows that $dS \cdot S^{-1}$ is skew-symmetric. Since S is orthogonal and ω is skew-symmetric,

$${}^t(S \cdot \omega \cdot S^{-1}) = {}^tS^{-1} \cdot {}^t\omega \cdot {}^tS = S \cdot {}^t\omega \cdot S^{-1} = -S \cdot \omega \cdot S^{-1}, \quad (3.2.29)$$

which shows that $S \cdot \omega \cdot S^{-1}$ is skew-symmetric. We set

$$dS \cdot S^{-1} = \begin{bmatrix} 0 & \alpha \\ -\alpha & 0 \end{bmatrix}, \qquad S \cdot \omega \cdot S^{-1} = \begin{bmatrix} 0 & \beta \\ -\beta & 0 \end{bmatrix}. \quad (3.2.30)$$

Since both α and β are 1-forms, we have

$$\begin{aligned} dS \cdot S^{-1} \wedge dS \cdot S^{-1} &= \begin{bmatrix} 0 & \alpha \\ -\alpha & 0 \end{bmatrix} \wedge \begin{bmatrix} 0 & \alpha \\ -\alpha & 0 \end{bmatrix} \\ &= \begin{bmatrix} -\alpha \wedge \alpha & 0 \\ 0 & -\alpha \wedge \alpha \end{bmatrix} \\ &= \begin{bmatrix} 0 & 0 \\ 0 & 0 \end{bmatrix}, \end{aligned} \quad (3.2.31)$$

and

$$\begin{aligned} & S \cdot \omega \cdot S^{-1} \wedge dS \cdot S^{-1} + dS \cdot S^{-1} \wedge S \cdot \omega \cdot S^{-1} \\ &= \begin{bmatrix} 0 & \beta \\ -\beta & 0 \end{bmatrix} \wedge \begin{bmatrix} 0 & \alpha \\ -\alpha & 0 \end{bmatrix} + \begin{bmatrix} 0 & \alpha \\ -\alpha & 0 \end{bmatrix} \wedge \begin{bmatrix} 0 & \beta \\ -\beta & 0 \end{bmatrix} \\ &= \begin{bmatrix} -\beta \wedge \alpha & 0 \\ 0 & -\beta \wedge \alpha \end{bmatrix} + \begin{bmatrix} -\alpha \wedge \beta & 0 \\ 0 & -\alpha \wedge \beta \end{bmatrix} \\ &= \begin{bmatrix} 0 & 0 \\ 0 & 0 \end{bmatrix}. \end{aligned} \quad (3.2.32)$$

Hence, (3.2.27) becomes as simple as

$$d\bar{\omega} = S \cdot d\omega \cdot S^{-1}. \quad (3.2.33)$$

Now, by substituting $d\omega_2{}^1 = K\theta^1 \wedge \theta^2$ and $d\bar{\omega}_2{}^1 = \bar{K}\bar{\theta}^1 \wedge \bar{\theta}^2$ into (3.2.33), we obtain

$$\begin{bmatrix} 0 & \bar{K}\bar{\theta}^1 \wedge \bar{\theta}^2 \\ -\bar{K}\bar{\theta}^1 \wedge \bar{\theta}^2 & 0 \end{bmatrix} = \begin{bmatrix} s_1{}^1 & s_2{}^1 \\ s_1{}^2 & s_2{}^2 \end{bmatrix} \begin{bmatrix} 0 & K\theta^1 \wedge \theta^2 \\ -K\theta^1 \wedge \theta^2 & 0 \end{bmatrix} \begin{bmatrix} s_1{}^1 & s_1{}^2 \\ s_2{}^1 & s_2{}^2 \end{bmatrix}. \quad (3.2.34)$$

Multiply out the right-hand side and compare the result with the left-hand side. Then we get

$$\bar{K}\bar{\theta}^1 \wedge \bar{\theta}^2 = ({s_1}^1{s_2}^2 - {s_2}^1{s_1}^2)\, K\theta^1 \wedge \theta^2. \tag{3.2.35}$$

On the other hand, from (3.2.10) we have

$$\bar{\theta}^1 \wedge \bar{\theta}^2 = ({s_1}^1{s_2}^2 - {s_2}^1{s_1}^2)\, \theta^1 \wedge \theta^2, \tag{3.2.36}$$

which, together with (3.2.25), gives

$$\bar{K} = K. \tag{3.2.37}$$

This shows that the Gaussian curvature K depends only on the Riemannian metric ds^2, not on the choice of (θ^1, θ^2) satisfying (3.2.1). In particular, this proves that although the Gaussian curvature of a surface in the space was defined in terms of the second fundamental form **II**, it actually depends only on the first fundamental form **I**. On discovering this fact, Gauss called it *Theorema egregium* (remarkable theorem).

As we proved by calculation in Example 2.3.1, a cylinder has the Gaussian curvature $K = 0$. The fact that its first fundamental form agrees with that of the plane is consistent with Gauss' Theorema egregium. On the other hand, the same example shows that the mean curvature H is not determined by the first fundamental form.

Now that we know at least theoretically that the Gaussian curvature K is determined by the Riemannian metric ds^2, we naturally face with the problem of explicitly expressing K in terms of ds^2. However, without the help of tensor analysis the necessary calculation would be very complicated. Hence we consider here only the case where ds^2 has the following special form:

$$ds^2 = E\, du du + G\, dv dv. \tag{3.2.38}$$

In this case, as θ^1, θ^2 we can take the following:

$$\theta^1 = \sqrt{E}\, du, \qquad \theta^2 = \sqrt{G}\, dv. \tag{3.2.39}$$

Then by (3.2.6) and (3.2.7),

$${\omega_2}^1 = \frac{1}{\sqrt{G}} \frac{\partial \sqrt{E}}{\partial v} du - \frac{1}{\sqrt{E}} \frac{\partial \sqrt{G}}{\partial u} dv. \tag{3.2.40}$$

Applying (3.2.9) to this, we obtain the following formula:

$$K = -\frac{1}{\sqrt{EG}} \left[\frac{\partial}{\partial u} \left(\frac{1}{\sqrt{E}} \frac{\partial \sqrt{G}}{\partial u} \right) + \frac{\partial}{\partial v} \left(\frac{1}{\sqrt{G}} \frac{\partial \sqrt{E}}{\partial v} \right) \right]. \tag{3.2.41}$$

In applications, the important is the following special case of (3.2.38):

$$ds^2 = E\, (du du + dv dv). \tag{3.2.42}$$

In this case, (3.2.41) is reduced to

$$K = -\frac{1}{E} \left(\frac{\partial^2}{\partial u^2} + \frac{\partial^2}{\partial v^2} \right) \log \sqrt{E} = -\frac{1}{2E} \left(\frac{\partial^2}{\partial u^2} + \frac{\partial^2}{\partial v^2} \right) \log E. \tag{3.2.43}$$

In both Example 3.1.1 and Example 3.1.2 of the preceding section, the Riemannian metrics are of the form (3.2.42). We can easily verify that both examples have the Gaussian curvature

$$K = -1. \tag{3.2.44}$$

The fact that the these two examples have the same Gaussian curvature is expected as a consequence of the existence of an isometric correspondence (3.1.13) and (3.1.15) between the two Riemannian metrics.

When a given Riemannian metric on a surface is of the form (3.2.42) in coordinates (u, v), we call (u, v) an **isothermal coordinate system**. It is known that for a given Riemannian metric there is always, locally, an isothermal coordinate system, but the proof of this fact is not easy. In Problem 3.2.2, we prove the existence of isothermal coordinate system in a few special cases. In Section 5.3 of Chapter 5, we shall prove the existence of an isothermal coordinate system for minimal surfaces. An isothermal coordinate system is essentially the same as a complex coordinate system, but we will not go into this question. The reader who has studied complex function theory will see the relationship between Problem 3.2.1 and the Cauchy-Riemann equations.

Problem 3.2.1 Given a surface with a Riemannian metric, consider two coordinate systems (u, v) and (x, y). If one of them is an isothermal coordinate system, then a necessary and sufficient condition for the other to be also isothermal is that (u, v) and (x, y) satisfy the following differential equations:

$$\frac{\partial u}{\partial x} = \frac{\partial v}{\partial y}, \quad \frac{\partial u}{\partial y} = -\frac{\partial v}{\partial x},$$

or

$$\frac{\partial u}{\partial x} = -\frac{\partial v}{\partial y}, \quad \frac{\partial u}{\partial y} = \frac{\partial v}{\partial x}.$$

Problem 3.2.2 Find an isothermal coordinate system (ξ, η) for the first fundamental form (see Example 2.3.6)

$$(f'(u)^2 + g'(u)^2)\, dudu + f(u)^2\, dvdv$$

of a surface of revolution. Namely, find a coordinate change $\xi = \xi(u, v)$, $\eta = \eta(u, v)$ which transforms the metric above to $\lambda\, d\xi d\xi + \lambda\, d\eta d\eta$.

In particular, find (ξ, η) explicitly for the metric

$$dudu + (u^2 + a^2)\, dvdv$$

of the catenoid and the right helicoid obtained in Problem 3.1.1.

Problem 3.2.3 When a Riemannian metric is of the form

$$ds^2 = (U + V)(dudu + dvdv),$$

where U is a function of u only while V is a function of v only, find its structure equations and the Gaussian curvature.

Problem 3.2.4 For a Riemannian metric

$$ds^2 = \frac{dudu - 4v\,dudv + 4u\,dvdv}{4(u - v^2)} \qquad (u > v^2),$$

find the structure equations and the Gaussian curvature.

3.3 Vector Fields

Since the time we learned calculus, we have been using the idea of tangent vectors of curves and surfaces in the space with geometric and intuitive understanding. However, when disregarding the fact that it is in the space and focusing on the surface itself, makes it necessary to think over what tangent vectors were.

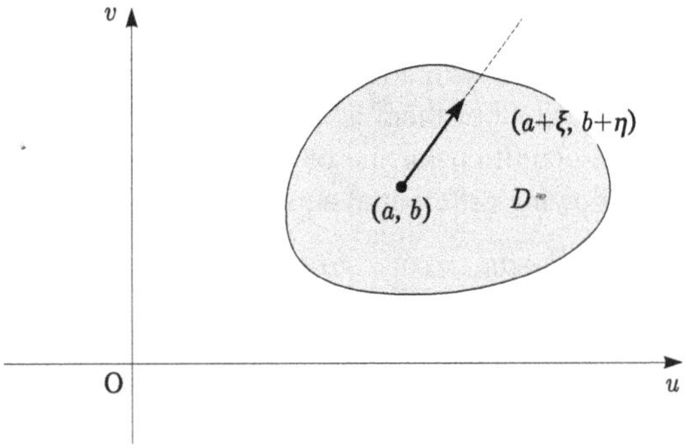

Fig. 3.3.1

Let us consider this question for a domain D in the (u, v)-plane which can be identified with a part of a general surface. In calculus, for a vector as in Figure 3.3.1, we considered directional derivatives. If f is a differentiable function defined near a point (a, b) in domain D, its derivative in the direction of the vector in Figure 3.3.1 is given by

$$\left(\xi \frac{\partial f}{\partial u} + \eta \frac{\partial f}{\partial v}\right)_{(u,v)=(a,b)}. \tag{3.3.1}$$

Thus, we regard the differential operator

$$\left(\xi \frac{\partial}{\partial u} + \eta \frac{\partial}{\partial v}\right)_{(u,v)=(a,b)} \tag{3.3.2}$$

itself as a **tangent vector**. (There are times when $(u, v) = (a, b)$ stands for the origin of the vector, we write simply (a, b) or completely omit it.)

Now, given a curve $(u(t), v(t))$ in the region D, differentiation of a function $f(u, v)$ on D along the curve is given by

$$\begin{aligned}\frac{df(u(t),v(t))}{dt} &= \frac{du}{dt}\frac{\partial f}{\partial u} + \frac{dv}{dt}\frac{\partial f}{\partial v} \\ &= \left(\frac{du}{dt}\frac{\partial}{\partial u} + \frac{dv}{dt}\frac{\partial}{\partial v}\right) f,\end{aligned} \tag{3.3.3}$$

proving that the tangent vector of the curve is nothing but

$$\frac{du}{dt}\frac{\partial}{\partial u} + \frac{dv}{dt}\frac{\partial}{\partial v}.$$

When assigning a tangent vector to each point on D, we call it a **vector field** on D. If ξ and η are functions on D,

$$X = \xi\frac{\partial}{\partial u} + \eta\frac{\partial}{\partial v} \tag{3.3.4}$$

gives a tangent vector at each point of D, thus representing a vector field on D. Without explicitly stating we assume that ξ and η are usually differentiable several times. Given a function f and a tangent vector field X on D, we have

$$Xf = \xi\frac{\partial f}{\partial u} + \eta\frac{\partial f}{\partial v}, \tag{3.3.5}$$

which is also a function on D.

Given a tangent vector field $X = \xi\frac{\partial}{\partial u} + \eta\frac{\partial}{\partial v}$ and a differential 1-form $\varphi = f du + g dv$, we define the product of φ and X by

$$\varphi(X) = f\xi + g\eta. \tag{3.3.6}$$

Sometimes we write $\langle\varphi, X\rangle$ instead of $\varphi(X)$. From this, for each point on D, the vector space of tangent vectors and that of differential 1-forms are dual spaces to each other. (Readers who have not learned the concept of dual space may ignore this word.)

Calculating the differential 1-form df from the function f and applying (3.3.6), we can easily obtain

$$\langle df, X\rangle = Xf. \tag{3.3.7}$$

Next, to see how the coordinate transformations affects the expression of a vector field, let us write (u^1, u^2) instead of (u, v) and (ξ^1, ξ^2) instead of (ξ, η). Consider another coordinate system (t^1, t^2), and the coordinate transformation

$$t^i = t^i(u^1, u^2) \qquad (i = 1, 2). \tag{3.3.8}$$

From

$$\frac{\partial}{\partial u^i} = \sum_{j=1}^{2}\frac{\partial t^j}{\partial u^i}\frac{\partial}{\partial t^j} \tag{3.3.9}$$

we have

$$X = \sum_{i=1}^{2}\xi^i\frac{\partial}{\partial u^i} = \sum_{j=1}^{2}\left(\sum_{i=1}^{2}\xi^i\frac{\partial t^j}{\partial u^i}\right)\frac{\partial}{\partial t^j}. \tag{3.3.10}$$

On the other hand, in (2.5.24), we proved that the differential 1-form $\varphi = \sum_{i=1}^{2} f_i \, du^i$ becomes

$$\varphi = \sum_{i=1}^{2} f_i \, du^i = \sum_{j=1}^{2} \left(\sum_{i=1}^{2} f_i \frac{\partial u^i}{\partial t^j} \right) dt^j. \tag{3.3.11}$$

Since we can easily prove that

$$\sum_{j=1}^{2} \left(\sum_{k=1}^{2} f_k \frac{\partial u^k}{\partial t^j} \right) \left(\sum_{i=1}^{2} \xi^i \frac{\partial t^j}{\partial u^i} \right) = \sum_{i=1}^{2} f_i \, \xi^i, \tag{3.3.12}$$

we see that the $\varphi(X)$ defined by (3.3.6) does not depend on the coordinate system.

Now, in general, when θ^1 and θ^2 are linearly independent differential 1-forms, vector fields $\boldsymbol{e}_1, \boldsymbol{e}_2$ such that

$$\theta^1(\boldsymbol{e}_1) = \theta^2(\boldsymbol{e}_2) = 1, \quad \theta^1(\boldsymbol{e}_2) = \theta^2(\boldsymbol{e}_1) = 0 \tag{3.3.13}$$

are uniquely determined. This is an elementary algebraic problem; if we set

$$\begin{cases} \theta^1 = a_1{}^1 \, du + a_2{}^1 \, dv, \\ \theta^2 = a_1{}^2 \, du + a_2{}^2 \, dv, \end{cases} \qquad \begin{cases} \boldsymbol{e}_1 = b_1{}^1 \dfrac{\partial}{\partial u} + b_1{}^2 \dfrac{\partial}{\partial v}, \\ \boldsymbol{e}_2 = b_2{}^1 \dfrac{\partial}{\partial u} + b_2{}^2 \dfrac{\partial}{\partial v}, \end{cases} \tag{3.3.14}$$

then (3.3.13) is nothing but the following matrix equation,

$$\begin{bmatrix} a_1{}^1 & a_2{}^1 \\ a_1{}^2 & a_2{}^2 \end{bmatrix} \begin{bmatrix} b_1{}^1 & b_2{}^1 \\ b_1{}^2 & b_2{}^2 \end{bmatrix} = \begin{bmatrix} 1 & 0 \\ 0 & 1 \end{bmatrix}. \tag{3.3.15}$$

Conversely, given linearly independent vector fields $\boldsymbol{e}_1, \boldsymbol{e}_2$, it is also clear that 1-forms θ^1, θ^2 satisfying (3.3.14) are uniquely determined. Readers who know the concept of dual space may know that $\boldsymbol{e}_1, \boldsymbol{e}_2$ and θ^1, θ^2 satisfying (3.3.13) are called the **dual base** of each other.

In Section 3.2 we expressed a Riemannian metric ds^2 as $\theta^1\theta^1 + \theta^2\theta^2$. Let $\boldsymbol{e}_1, \boldsymbol{e}_2$ be the vector fields corresponding to 1-forms θ^1, θ^2 (the dual base). For two tangent vectors X, Y written in the form

$$X = \xi^1 \boldsymbol{e}_1 + \xi^2 \boldsymbol{e}_2, \qquad Y = \eta^1 \boldsymbol{e}_1 + \eta^2 \boldsymbol{e}_2, \tag{3.3.16}$$

we define an inner product $\langle X, Y \rangle$ and a length $|X|$ as

$$\begin{aligned} \langle X, Y \rangle &= \xi^1 \eta^1 + \xi^2 \eta^2, \\ |X| &= \langle X, Y \rangle^{\frac{1}{2}}. \end{aligned} \tag{3.3.17}$$

In other words we define an inner product in each tangent space of D in such a way that $\boldsymbol{e}_1, \boldsymbol{e}_2$ become orthonormal (i.e., $|\boldsymbol{e}_1| = |\boldsymbol{e}_2| = 1, \langle \boldsymbol{e}_1, \boldsymbol{e}_2 \rangle = 0$). In Section 3.1 we defined the Riemannian metric formally as a generalization of the first fundamental form, but the essential point is that it defines an inner product in each tangent space.

Now, let us use different 1-forms $\bar{\theta}^1$, $\bar{\theta}^2$ to write ds^2 as

$$ds^2 = \bar{\theta}^1\,\bar{\theta}^1 + \bar{\theta}^2\,\bar{\theta}^2,$$

and let $\bar{\boldsymbol{e}}_1$, $\bar{\boldsymbol{e}}_2$ be the corresponding vector fields. Then, we have to check that the inner product and the length defined by $\boldsymbol{e}_1$, $\boldsymbol{e}_2$ coincide with those defined by $\bar{\boldsymbol{e}}_1$, $\bar{\boldsymbol{e}}_2$. Let

$$\begin{cases} \bar{\theta}^1 & = {s_1}^1\,\theta^1 + {s_2}^1\,\theta^2, \\ \bar{\theta}^2 & = {s_1}^2\,\theta^1 + {s_2}^2\,\theta^2, \end{cases} \qquad \begin{cases} \bar{\boldsymbol{e}}_1 & = {t_1}^1\,\boldsymbol{e}_1 + {t_1}^2\,\boldsymbol{e}_2, \\ \bar{\boldsymbol{e}}_2 & = {t_2}^1\,\boldsymbol{e}_1 + {t_2}^2\,\boldsymbol{e}_2. \end{cases} \tag{3.3.18}$$

Using the matrix notation

$$\begin{aligned} \theta &= \begin{bmatrix} \theta^1 \\ \theta^2 \end{bmatrix}, & \boldsymbol{e} &= (\boldsymbol{e}_1, \boldsymbol{e}_2), \\ S &= \begin{bmatrix} {s_1}^1 & {s_2}^1 \\ {s_1}^2 & {s_2}^2 \end{bmatrix}, & T &= \begin{bmatrix} {t_1}^1 & {t_2}^1 \\ {t_1}^2 & {t_2}^2 \end{bmatrix}, \end{aligned} \tag{3.3.19}$$

this can be simply rewritten as

$$\bar{\theta} = S\,\theta, \qquad \bar{\boldsymbol{e}} = \boldsymbol{e}\,T. \tag{3.3.20}$$

As in the case of (3.3.15), we have

$$T\,S = I, \qquad \text{where } I \text{ is the unit matrix.} \tag{3.3.21}$$

On the other hand, from $\bar{\theta}^1\bar{\theta}^1 + \bar{\theta}^2\bar{\theta}^2 = \theta^1\theta^1 + \theta^2\theta^2$ we get ${}^t S \cdot S = I$, which shows that S is an orthogonal matrix. Thus T is also an orthogonal matrix, and $\bar{\boldsymbol{e}}_1$, $\bar{\boldsymbol{e}}_2$ form an orthonormal system. From the above, we see that the inner product and the length of vectors are determined by only the Riemannian metric and do not depend on the choice of θ^1, θ^2. Finally, we note that (3.3.21) can also be written as

$$\bar{\boldsymbol{e}}\,\bar{\theta} = \boldsymbol{e}\,\theta. \tag{3.3.22}$$

Problem 3.3.1 For given vector fields $X = \sum_{i=1}^{2} \xi^i \frac{\partial}{\partial u^i}$ and $Y = \sum_{i=1}^{2} \eta^i \frac{\partial}{\partial u^i}$, define a third vector field $[X,Y]$ by

$$[X,Y]f = X(Yf) - Y(Xf).$$

(In the above equation we are considering these vector fields as differential operators acting on functions.) Express this $[X,Y]$ using ξ^i, η^i.

Next, check the Jacobi identity:

$$[[X,Y],Z] + [[Y,Z],X] + [[Z,X],Y] = 0$$

for three vector fields X, Y, Z.

Problem 3.3.2 For a differential 2-form $\varphi = f du^1 \wedge du^2$ and two vector fields $X = \sum_{i=1}^{2} \xi^i \frac{\partial}{\partial u^i}$, $Y = \sum_{i=1}^{2} \eta^i \frac{\partial}{\partial u^i}$, define

$$\varphi(X,Y) = f(\xi^1\eta^2 - \xi^2\eta^1).$$

Check the following.

(a) If φ and ψ are 1-forms, then

$$(\varphi \wedge \psi)(X,Y) = \varphi(X)\psi(Y) - \psi(X)\varphi(Y).$$

(b) If φ is a 1-form, then

$$(d\varphi)(X,Y) = X(\varphi(Y)) - Y(\varphi(X)) - \varphi([X,Y]).$$

3.4 Covariant Derivatives and Parallel Translations

First we shall start with the case of a surface $\boldsymbol{p}(u,v)$ in the space. Let a tangent vector $X(t)$ be given at each point on curve $\boldsymbol{p}(t) = \boldsymbol{p}(u(t),v(t))$ on this surface. (We call this a vector field along the curve.) Since $X(t)$ is a vector in the space, it has three components and by differentiating every component for t, we define $X'(t)$ (or $\frac{dX}{dt}$). $X'(t)$ also has three components, so we can assume that it is also a vector in the space, but it is not necessarily tangent to the surface $\boldsymbol{p}(u,v)$. We write $X'(t)$ as a sum of a tangent vector and normal vector of this surface:

$$X' = \frac{DX}{dt} + A_X \tag{3.4.1}$$

with the tangent vector $\frac{DX}{dt}$ and the normal vector A_X. If we focus only on the surface and ignore the surrounding space, only the tangential component $\frac{DX}{dt}$ of X' is meaningful. We call $\frac{DX}{dt}$ the **covariant derivative** of X along the curve $\boldsymbol{p}(t)$. What is important is that $\frac{DX}{dt}$ depends only on the first fundamental form of the surface and not on the relative position of the surface in the space, in other words, it is independent of the second fundamental form. In order to check this fact, taking an orthonormal system $\boldsymbol{e}_1, \boldsymbol{e}_2$, set $\boldsymbol{e}_3 = \boldsymbol{e}_1 \times \boldsymbol{e}_2$ as in Section 2.4 of Chapter 2. We write

$$X = \xi^1 \boldsymbol{e}_1 + \xi^2 \boldsymbol{e}_2 \tag{3.4.2}$$

and calculate X'. Using (2.4.12) we have

$$\begin{aligned}\frac{dX}{dt} &= \left(\frac{d\xi^1}{dt} + \xi^1 \frac{{\omega_1}^1}{dt} + \xi^2 \frac{{\omega_2}^1}{dt}\right)\boldsymbol{e}_1 \\ &+ \left(\frac{d\xi^2}{dt} + \xi^1 \frac{{\omega_1}^2}{dt} + \xi^2 \frac{{\omega_2}^2}{dt}\right)\boldsymbol{e}_2 + \left(\xi^1 \frac{{\omega_1}^3}{dt} + \xi^2 \frac{{\omega_2}^3}{dt}\right)\boldsymbol{e}_3.\end{aligned} \tag{3.4.3}$$

(The curious notation $\frac{{\omega_j}^i}{dt}$ we used here should be understood as follows. For a differential form $\omega = a\,du + b\,dv$ let $\frac{\omega}{dt}$ mean $a\,\frac{du}{dt} + b\,\frac{dv}{dt}$.) Noting that ${\omega_1}^1 = {\omega_2}^2 = 0$ ((2.4.14) of Chapter 2), by separating (3.4.3) into the tangential components and the normal components we write

$$\frac{DX}{dt} = \left(\frac{d\xi^1}{dt} + \xi^2 \frac{{\omega_2}^1}{dt}\right)\boldsymbol{e}_1 + \left(\frac{d\xi^2}{dt} + \xi^1 \frac{{\omega_1}^2}{dt}\right)\boldsymbol{e}_2, \tag{3.4.4}$$

$$A_X = \left(\xi^1 \frac{{\omega_1}^3}{dt} + \xi^2 \frac{{\omega_2}^3}{dt}\right) e_3. \tag{3.4.5}$$

We can see that in (3.4.4) only the components ${\omega_2}^1, {\omega_1}^2$ of the connection form are used for $\frac{DX}{dt}$, and not ${\omega_1}^3, {\omega_2}^3$ which define the second fundamental form. The forms ${\omega_2}^1, {\omega_1}^2 (= -{\omega_2}^1)$ defined in (3.2.8) were uniquely determined by the Riemannian metric ds^2 and $\boldsymbol{e}_1, \boldsymbol{e}_2$ (or θ^1, θ^2). Thus $\frac{DX}{dt}$ defined in (3.4.4) is determined by ds^2 and $\boldsymbol{e}_1, \boldsymbol{e}_2$, but in (3.4.10) we will prove that it does not depend on the choice of $\boldsymbol{e}_1$, $\boldsymbol{e}_2$.

When $\frac{DX}{dt} \equiv \boldsymbol{0}$, we say that X is **parallel** along the curve $(u(t), v(t))$. When the parameter t moves from a to b, a vector $X(a)$ given at the point $(u(a), v(a))$ can be uniquely moved along the curve so that $\frac{DX}{dt} \equiv \boldsymbol{0}$. From (3.4.4) we know that this is the same as solving the ordinary differential equation

$$\begin{cases} \dfrac{d\xi^1}{dt} + \xi^2 \dfrac{{\omega_2}^1}{dt} = 0, \\ \dfrac{d\xi^2}{dt} + \xi^1 \dfrac{{\omega_1}^2}{dt} = 0, \end{cases} \tag{3.4.6}$$

or

$$\frac{d\xi^i}{dt} + \sum_{j=1}^{2} \frac{{\omega_j}^i}{dt} \xi^j = 0$$

by unifying, under the initial condition $\xi^i(a) = \alpha^i$ $(i = 1, 2)$. (For the existence and the uniqueness of solution, see a textbook on ordinary differential equations and also the appendix on differential equations at the end of this book.)

To summarize, given a Riemannian metric ds^2 on a domain D in the plane, writing $ds^2 = \theta^1\theta^1 + \theta^2\theta^2$ and taking vector fields $\boldsymbol{e}_1, \boldsymbol{e}_2$ corresponding to θ^1, θ^2, we define by (3.4.4) the covariant derivative $\frac{DX}{dt}$ of a vector field $X = \xi^1\boldsymbol{e}_1 + \xi^2\boldsymbol{e}_2$ along a curve $(u(t), v(t))$. When $\frac{DX}{dt} = \boldsymbol{0}$, i.e., (3.4.6) holds, we say that X is **parallel** along the curve. Then we need to check that $\frac{DX}{dt}$ does not depend on the choice of θ^1, θ^2 in $ds^2 = \theta^1\theta^1 + \theta^2\theta^2$. To do so, as in (3.3.19), we use the matrix notation

$$\xi = \begin{bmatrix} \xi^1 \\ \xi^2 \end{bmatrix}, \qquad \boldsymbol{e} = (\boldsymbol{e}_1, \boldsymbol{e}_2), \qquad \omega = \begin{bmatrix} {\omega_1}^1 & {\omega_2}^1 \\ {\omega_1}^2 & {\omega_2}^2 \end{bmatrix}, \tag{3.4.7}$$

where ${\omega_1}^1 = {\omega_2}^2 = 0$, to rewrite (3.4.4) as

$$\frac{DX}{dt} = \boldsymbol{e}\left(\frac{d\xi}{dt} + \frac{\omega}{dt}\xi\right). \tag{3.4.8}$$

Using

$$\bar{\boldsymbol{e}} = \boldsymbol{e}S^{-1}, \qquad \bar{\xi} = S\xi, \qquad \bar{\omega} = S\omega S^{-1} - dS \cdot S^{-1}, \tag{3.4.9}$$

we have

$$\bar{\boldsymbol{e}}\left(\frac{d\bar{\xi}}{dt} + \frac{\bar{\omega}}{dt}\bar{\xi}\right) = \boldsymbol{e}S^{-1}\left(\frac{dS}{dt}\xi + S\frac{d\xi}{dt} + S\frac{\omega}{dt}S^{-1}S\xi - \frac{dS}{dt}S^{-1}S\xi\right).$$

Simplifying the right-hand side,

$$\bar{e}\left(\frac{d\bar{\xi}}{dt}+\frac{\bar{\omega}}{dt}\bar{\xi}\right)=e\left(\frac{d\xi}{dt}+\frac{\omega}{dt}\xi\right). \tag{3.4.10}$$

Next we shall prove that the inner product and the length of vectors defined in the previous section are invariant by parallel translation. Let

$$X=\sum_{i=1}^{2}\xi^i\,e_i,\qquad Y=\sum_{i=1}^{2}\eta^i\,e_i \tag{3.4.11}$$

be two vector fields parallel along a curve $(u(t), v(t))$. Then we have

$$\begin{aligned}\frac{d}{dt}\langle X,Y\rangle &= \frac{d}{dt}\left(\sum_{i=1}^{2}\xi^i\,\eta^i\right)=\sum_{i=1}^{2}\frac{d\xi^i}{dt}\eta^i+\sum_{i=1}^{2}\xi^i\,\frac{d\eta^i}{dt}\\ &=\sum_{i=1}^{2}\frac{d\xi^i}{dt}\eta^i+\sum_{i,j=1}^{2}\frac{\omega_j{}^i}{dt}\xi^j\,\eta^i+\sum_{i=1}^{2}\xi^i\,\frac{d\eta^i}{dt}+\sum_{i,j=1}^{2}\xi^i\,\frac{\omega_j{}^i}{dt}\eta^j\\ &=\left\langle\frac{DX}{dt},Y\right\rangle+\left\langle X,\frac{DX}{dt}\right\rangle=0.\end{aligned} \tag{3.4.12}$$

(Since $\omega_j{}^i=-\omega_i{}^j$, the second and fourth terms in the middle cancel out each other, giving the third equality. Using (3.4.4) we obtain the fourth equality, and the last equality is a result of the assumption that X and Y are parallel vector fields, in other words, $\frac{DX}{dt}=\frac{DY}{dt}=\mathbf{0}$.) Thus, $\langle X,Y\rangle$ is invariant along the curve $(u(t), v(t))$. The length of X is also invariant since it is defined as $\langle X,X\rangle^{\frac{1}{2}}$.

Thus, the concept of parallel translation resembles that in Euclidean geometry, however there is one big difference.

In Euclidean geometry, it is meaningful to say whether vectors at two separate points are parallel or not. However on surfaces, we can only discuss if vectors at two separate points are parallel **along a given curve** or not. Thus, given two curves connecting the two points, vectors may be parallel along one curve, but not parallel along the other. In order to understand this, let us examine the parallel translation along a small circle on a sphere.

Example 3.4.1 Parallel translation along a small circle on a sphere.

As in Figure 3.4.1, set parameters u, v on the sphere with radius a so that a point (x, y, z) on the sphere is given by

$$x=a\cos u\cos v,\quad y=a\cos u\sin v,\quad z=a\sin u. \tag{3.4.13}$$

Excluding both poles $\left(\text{i.e., } u=\frac{\pi}{2},-\frac{\pi}{2}\right)$, we define e_1, e_2 as follows. Let e_1 be the unit tangent vector of the latitude line for a fixed u and moving v; let e_2 be the unit tangent vector for the meridian for a fixed v and moving u. Thus,

$$\begin{aligned}e_1&=(-\sin v,\ \cos v,\ 0),\\ e_2&=(-\sin u\cos v,\ -\sin u\sin v,\ \cos u),\end{aligned} \tag{3.4.14}$$

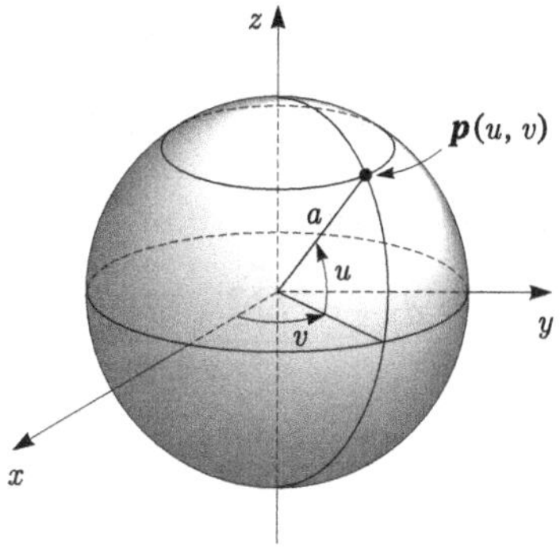

Fig. 3.4.1

and the normal vector $e_3 = e_1 \times e_2$ is given by

$$e_3 = (\cos v \cos u, \sin v \cos u, \sin u). \tag{3.4.15}$$

We express de_1, de_2, de_3 as linear combinations of e_1, e_2, e_3:

$$\begin{cases} de_1 = \sin u \, dv \, e_2 - \cos u \, dv \, e_3, \\ de_2 = -\sin u \, dv \, e_1 - du \, e_3, \\ de_3 = \cos u \, dv \, e_1 + du \, e_2. \end{cases} \tag{3.4.16}$$

Using matrix notations, the connection form is given as

$$\begin{bmatrix} \omega_1{}^1 & \omega_2{}^1 & \omega_3{}^1 \\ \omega_1{}^2 & \omega_2{}^2 & \omega_3{}^2 \\ \omega_1{}^3 & \omega_2{}^3 & \omega_3{}^3 \end{bmatrix} = \begin{bmatrix} 0 & -\sin u \, dv & \cos u \, dv \\ \sin u \, dv & 0 & du \\ -\cos u \, dv & -du & 0 \end{bmatrix}. \tag{3.4.17}$$

Set

$$\lambda = \sin u \tag{3.4.18}$$

to determine a vector field $X = \xi^1 e_1 + \xi^2 e_2$ which is parallel along the latitude line obtained when fixing u. When v is used as the parameter of the curve, equation (3.4.6) is given by

$$\begin{cases} \dfrac{d\xi^1}{dv} - \lambda \xi^2 = 0, \\ \\ \dfrac{d\xi^2}{dv} + \lambda \xi^1 = 0. \end{cases} \tag{3.4.19}$$

The general solution of this differential equation is given as follows.

$$\begin{cases} \xi^1 = \alpha \sin \lambda v - \beta \cos \lambda v, \\ \xi^2 = \beta \sin \lambda v + \alpha \cos \lambda v. \end{cases} \tag{3.4.20}$$

Thus, we now know about the parallel translation along the latitude line. For example, take the vector e_1 ($\alpha = 0, \beta = -1$ in (3.4.20)) at $v = 0$. By parallel translating it along the latitude line in the positive direction from $v = 0$ to $v = \pi$, we obtain $(\cos \lambda\pi)\, \boldsymbol{e}_1 - (\sin \lambda\pi)\, \boldsymbol{e}_2$. For the same vector, we obtain $(\cos \lambda\pi)\, \boldsymbol{e}_1 + (\sin \lambda\pi)\, \boldsymbol{e}_2$ by parallel translating it along the latitude line in the negative direction to $v = -\pi$. ♦

Problem 3.4.1 Let U be the upper half $v > 0$ of the $w = u + iv$ plane. For the Poincaré metric

$$\frac{dw\, d\bar{w}}{v^2} = \frac{(du)^2 + (dv)^2}{v^2} \tag{3.4.21}$$

(see Example 3.1.2), consider an orthonormal system $\boldsymbol{e}_1 = v\, \frac{\partial}{\partial u}$, $\boldsymbol{e}_2 = v\, \frac{\partial}{\partial v}$.

(a) Determine $\omega_2{}^1$.
(b) Determine the parallel vector fields along a curve $u = a \cos t$, $v = a \sin t$ $(0 < t < \pi)$.
(c) Determine the parallel vector fields along a line $u = t$, $v = a$ $(0 < t < \pi)$.

Problem 3.4.2 Using Christoffel's symbols Γ^i_{jk} (see Problem 2.4.1), prove that the covariant derivative of a vector field $X = \sum_{i=1}^2 \xi^i \frac{\partial}{\partial u^i}$ along a curve $(u^1(t), u^2(t))$ is given by

$$\frac{DX}{dt} = \sum_{i=1}^{2} \left(\frac{d\xi^i}{dt} + \sum_{j,k=1}^{2} \Gamma^i_{jk}\, \xi^j \frac{du^k}{dt} \right) \frac{\partial}{\partial u^i}.$$

3.5 Geodesics

We will now review the concept of a **geodesic** for a surface in the space, which we defined in Section 2.2 of Chapter 2. For a curve $\boldsymbol{p}(s) = \boldsymbol{p}(u(s), v(s))$ on a surface $\boldsymbol{p}(u, v)$, set the parameter s in such a way that the length of the tangent vector $\boldsymbol{p}'(s)$ is 1. The acceleration vector $\boldsymbol{p}''(s)$ is generally not tangent to the surface, and can be written as

$$\boldsymbol{p}''(s) = \boldsymbol{k}_g + \boldsymbol{k}_n, \tag{3.5.1}$$

where $\boldsymbol{k}_g$ is the geodesic curvature vector tangent to the surface, and $\boldsymbol{k}_n$ is the normal curvature vector perpendicular to the surface. The curve $\boldsymbol{p}(s)$ is called a geodesic when $\boldsymbol{k}_g = \boldsymbol{0}$, i.e., $\boldsymbol{p}''(s)$ is perpendicular to the surface. In Chapter 2, we have studied the relationships between $\boldsymbol{k}_n$ and the second fundamental form (see (2.2.36)). In this section we will learn that the vector field $\boldsymbol{k}_g$ and geodesics are both determined by the first fundamental forms.

First of all, using an orthonormal frame $\boldsymbol{e}_1$, $\boldsymbol{e}_2$, $\boldsymbol{e}_3$, we express the unit tangent vector $\boldsymbol{p}'(s)$ as

$$\boldsymbol{p}'(s) = \sum_{i=1}^{2} \xi^i\, \boldsymbol{e}_i \tag{3.5.2}$$

and differentiating it, we obtain

$$
\begin{aligned}
p''(s) &= \sum_{i=1}^{2}\left(\frac{d\xi^i}{ds}\,e_i + \xi^i\,\frac{de_i}{ds}\right)\\
&= \sum_{i=1}^{2}\frac{d\xi^i}{ds}\,e_i + \sum_{i=1}^{2}\xi^i\sum_{j=1}^{3}\frac{\omega_i{}^j}{ds}\,e_j \qquad (3.5.3)\\
&= \sum_{i=1}^{2}\left(\frac{d\xi^i}{ds} + \sum_{j=1}^{2}\frac{\omega_j{}^i}{ds}\,\xi^j\right)e_i + \sum_{j=1}^{2}\frac{\omega_j{}^3}{ds}\,\xi^j\,e_3.
\end{aligned}
$$

Thus we have

$$
\begin{aligned}
k_g &= \sum_{i=1}^{2}\left(\frac{d\xi^i}{ds} + \sum_{j=1}^{2}\frac{\omega_j{}^i}{ds}\,\xi^j\right)e_i,\\
k_n &= \sum_{j=1}^{2}\frac{\omega_j{}^3}{ds}\,\xi^j\,e_3.
\end{aligned} \qquad (3.5.4)
$$

By setting

$$X = p'(s), \qquad (3.5.5)$$

and comparing (3.5.4) with (3.4.4) and (3.4.5), we can write it as

$$k_g = \frac{DX}{ds}, \qquad k_n = A_X. \qquad (3.5.6)$$

From this we see that k_g is the covariant derivative of the tangent vector field $p'(s)$ along the curve $p(s)$. Therefore it is now clear that k_g is determined only by the first fundamental form. Thus geodesics are given by $k_g = \mathbf{0}$, i.e.,

$$\frac{DX}{ds} = \mathbf{0}, \qquad \text{where } X = p'(s). \qquad (3.5.7)$$

Thus we know that when a given curve is a **geodesic**, its tangent vector field is parallel along the curve itself. From this characterization, the concept of a geodesic can be defined for surfaces which do not lie in the 3-dimensional Euclidean space.

Equation (3.5.7) is an ordinary differential equation of degree two (in Problem 3.5.2 this will be made clearer). When a starting point ($u_0 = u(0), v_0 = v(0)$) and a unit tangent vector $X_0 = \xi_0{}^1\,e_1 + \xi_0{}^2\,e_2$ at (u_0, v_0) are given as an initial condition, there exists a unique solution of (3.5.7) where this vector is the acceleration vector at the initial point, i.e., a geodesic $(u(s), v(s))$. For this fact, see a textbook on differential equations and also the appendix at the end of this book.

Now we will simplify (3.5.4) using the special frame e_1, e_2, e_3. Since $p'(s)$ is a unit vector, let

$$e_1(s) = p'(s), \qquad (3.5.8)$$

and let e_2 be a unit vector orthogonal to e_1 and tangent to the surface. (By the condition that the direction of rotation from e_1 to e_2 is the same as that of p_u to p_v,

e_2 is uniquely determined.) On the other hand, by differentiating $\boldsymbol{p}'(s) \cdot \boldsymbol{p}'(s) = 1$ we obtain $\boldsymbol{p}''(s) \cdot \boldsymbol{p}'(s) = 0$ and $\boldsymbol{p}''(s)$ is orthogonal to $\boldsymbol{p}'(s)$, thus $\boldsymbol{k}_g$ is a scalar multiple of $\boldsymbol{e}_2$, and $\boldsymbol{k}_n$ is a scalar multiple of $\boldsymbol{e}_3 = \boldsymbol{e}_1 \times \boldsymbol{e}_2$. Thus setting

$$\boldsymbol{k}_g = \kappa_g\, \boldsymbol{e}_2, \qquad \boldsymbol{k}_n = \kappa_n\, \boldsymbol{e}_3, \tag{3.5.9}$$

we call κ_g the **geodesic curvature**, and κ_n the **normal curvature** of $\boldsymbol{p}(s)$. Calculating κ_g using connection form (3.2.8) corresponding to the above $\boldsymbol{e}_1, \boldsymbol{e}_2, \boldsymbol{e}_3$ we obtain,

$$\begin{aligned} \kappa_g &= \kappa_g\, \boldsymbol{e}_2 \cdot \boldsymbol{e}_2 = \boldsymbol{k}_g \cdot \boldsymbol{e}_2 = \boldsymbol{p}'' \cdot \boldsymbol{e}_2 = \boldsymbol{e}_1' \cdot \boldsymbol{e}_2 \\ &= \frac{d\boldsymbol{e}_1}{ds} \cdot \boldsymbol{e}_2 = \frac{{\omega_1}^2}{ds}\, \boldsymbol{e}_2 \cdot \boldsymbol{e}_2 = \frac{{\omega_1}^2}{ds}. \end{aligned} \tag{3.5.10}$$

We can obtain this equation by setting $\xi^1 = 1$, $\xi^2 = 0$ in (3.5.4). From a similar calculation, we obtain

$$\kappa_n = \frac{{\omega_1}^3}{ds}. \tag{3.5.11}$$

Let us now understand geodesic curvature in a more geometric way. When we calculate κ_g at a point $\boldsymbol{p}_0 = \boldsymbol{p}(s_0)$, we consider a plane curve $\boldsymbol{q}(s)$ obtained by orthogonally projecting $\boldsymbol{p}(s)$ to the tangent plane at $\boldsymbol{p}_0$ for the surface $\boldsymbol{p}(u, v)$. By

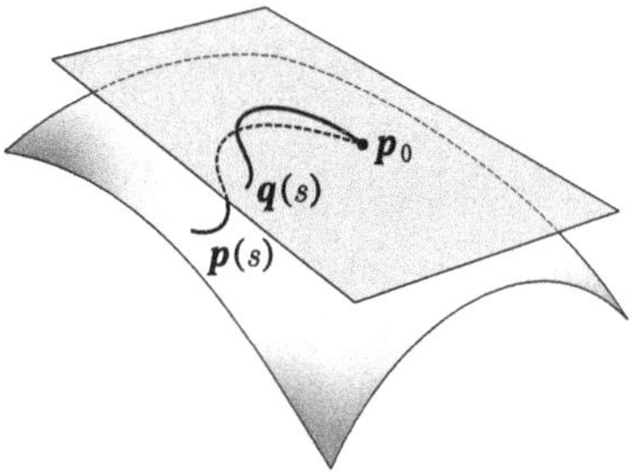

Fig. 3.5.1

taking a unit normal vector $\boldsymbol{e}$ to the surface at the point $\boldsymbol{p}_0$ we can write as follows

$$\boldsymbol{p}(s) = \boldsymbol{q}(s) + (\boldsymbol{p}(s) \cdot \boldsymbol{e})\, \boldsymbol{e}. \tag{3.5.12}$$

By differentiating twice, we get

$$\begin{cases} \boldsymbol{p}'(s_0) = \boldsymbol{q}'(s_0), & \text{since } \boldsymbol{p}'(s_0) \cdot \boldsymbol{e} = 0, \\ \boldsymbol{p}''(s_0) = \boldsymbol{q}''(s_0) + (\boldsymbol{p}''(s_0) \cdot \boldsymbol{e})\, \boldsymbol{e}. \end{cases} \tag{3.5.13}$$

Since $\boldsymbol{q}(s)$ is in the tangent plane at $\boldsymbol{p}_0$, $\boldsymbol{q}''(s)$ is also in the tangent plane. By comparing (3.5.13) and (3.5.1) we get

$$\boldsymbol{q}''(s_0) = \boldsymbol{k}_g(s_0), \qquad (\boldsymbol{p}''(s_0) \cdot \boldsymbol{e})\, \boldsymbol{e} = \boldsymbol{k}_n(s_0). \tag{3.5.14}$$

On the other hand, from (1.2.21) of Chapter 1 we can see that the length of $\boldsymbol{q}''(s_0)$ is the curvature of the plane curve $\boldsymbol{q}(s)$ at the point $\boldsymbol{q}(s_0)$. (The point about the above argument is that s is the length parameter for $\boldsymbol{p}(s)$, but for $\boldsymbol{q}(s)$ it is not necessarily the length parameter at places other than $s = s_0$, therefore in order to see that $|\boldsymbol{q}''(s_0)|$ is the curvature, we need to do some calculation. This calculation would be appropriate for exercise.)

Next, we shall explain a few examples of geodesics.

Example 3.5.1 In a plane, we can easily tell that the geodesic is nothing but a line. The idea of geodesics came from extending the idea of lines on planes to surfaces. ♦

Example 3.5.2 For any point in a sphere, when a great circle is orthogonally projected to the tangent plane at that point, it becomes a line. Therefore, the geodesic curvature

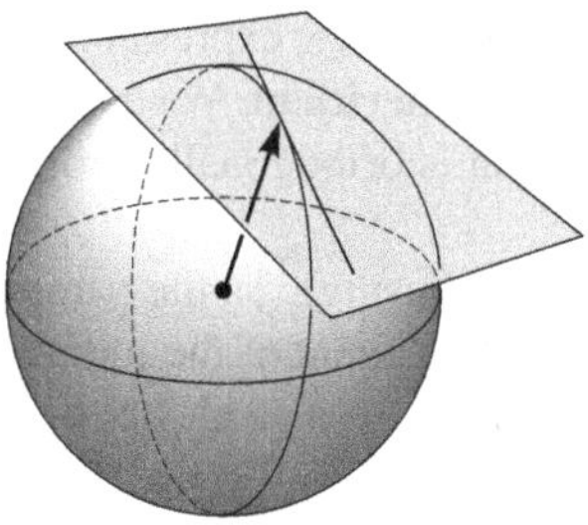

Fig. 3.5.2

κ_g of a great circle is zero (in the above argument, if we set a great circle as $\boldsymbol{p}(s)$, $\boldsymbol{q}(s)$ will be the line). Therefore, a great circle is a geodesic. Generally, using the fact that when a tangent vector is given on a surface there is only one geodesic tangent to it (we have already mentioned that we can prove this by using ordinary differential equations), we can tell that a geodesic of a sphere must be a great circle.

Instead of the above geometrical argument, let us now calculate κ_g of the latitude, using Example 3.4.1. Using the notation in Example 3.4.1, we fix u and move v. Since we had set up frames in such a way that $\boldsymbol{e}_1$ is the tangent vector field of the latitude, we can obtain κ_g using (3.5.10). From (3.4.17) we know that ${\omega_1}^2$ is $\sin u\, dv$, so $\kappa_g = \sin u \frac{dv}{ds}$. Since the radius of the sphere is a, the small circle given by the latitude will have radius $a \cos u$. Hence, when $s = av \cos u$, s will be the length parameter, giving

$$\kappa_g = \frac{1}{a} \tan u. \tag{3.5.15}$$

Here, when $u = 0$, $\kappa_g = 0$ showing that a great circle is a geodesic. ♦

Example 3.5.3 For an ellipsoid : $\frac{x^2}{a^2} + \frac{y^2}{b^2} + \frac{z^2}{c^2} = 1$, from the same reason as Example 3.5.2, we can also see that the three closed curves which are the intersections of the ellipsoid with the coordinate planes, are geodesics. In general, there is a theorem saying that “for any closed surface there are always at least three closed geodesics,” but this is a very difficult theorem. ♦

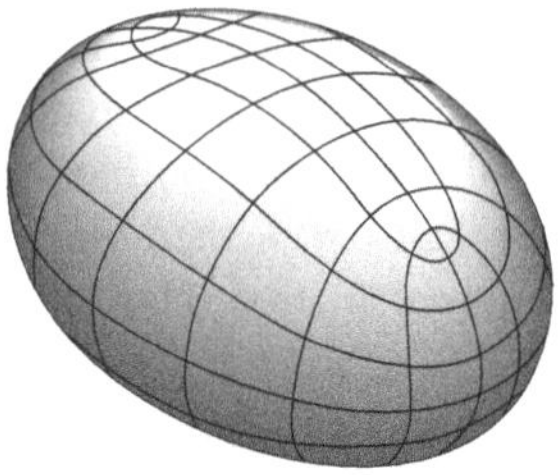

Fig. 3.5.3

Example 3.5.4 For a surface of revolution, it is difficult to determine all geodesics, but we can easily find special geodesics. Let $x = f(u)$, $z = g(u)$ be a curve on the (x, z)-plane. Revolving this curve around the z-axis with angle v, let $\boldsymbol{p}(u, v) = (f(u)\cos v, f(u)\sin v, g(u))$ be a parametric expression. Fixing v, we get a curve with parameter u. We call this curve the **meridian**. We can easily see that the meridian is a geodesic since when a meridian is orthogonally projected to the tangent plane at that point, it becomes a line.

On the other hand, fixing u, and using v as the parameter we get a **parallel**. For u where $f'(u) = 0$, i.e., where the gradient of the meridian is in the direction of the z-axis, the parallel corresponding to u is a geodesic. ♦

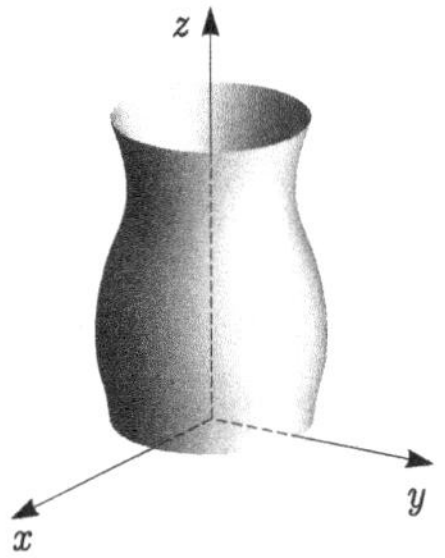

Fig. 3.5.4

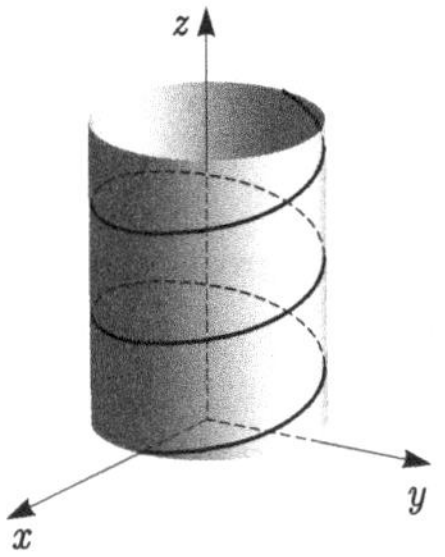

Fig. 3.5.5

Example 3.5.5 For a right circular cylinder, we cut it along a generating line and develop it on a plane. A curve on the cylinder is a geodesic if and only if it becomes a line when developed. Hence, a geodesic is either a **helix** or a generating line or a parallel. Beans growing along the helix, i.e., along a geodesic, might be the dispensation of nature. ♦

Example 3.5.6 Let us calculate the geodesics of the upper-half plane $U = \{(x, y)\,;\, y > 0\}$ for the Poincaré metric

$$ds^2 = \frac{(dx)^2 + (dy)^2}{y^2}. \tag{3.5.16}$$

Unlike Examples (3.5.2)–(3.5.5), it is not a surface in the space, thus we will solve it using the differential equation defining the geodesic. Set

$$\theta^1 = \frac{dx}{y}, \qquad \theta^2 = \frac{dy}{y}, \tag{3.5.17}$$

and we get $ds^2 = (\theta^1)^2 + (\theta^2)^2$. Since

$$d\theta^1 = \frac{dx \wedge dy}{y^2} = \frac{dx}{y} \wedge \theta^2, \qquad d\theta^2 = 0 = -\frac{dx}{y} \wedge \theta^1, \tag{3.5.18}$$

by setting ${\omega_2}^1 = -\frac{dx}{y} = -{\omega_1}^2$ we get

$$d\theta^1 = -{\omega_2}^1 \wedge \theta^2, \qquad d\theta^2 = -{\omega_1}^2 \wedge \theta^1, \tag{3.5.19}$$

thus we now have the necessary structure equations. Next, let $(x(s), y(s))$ be the geodesic and let us express its tangent vector field

$$x' \frac{\partial}{\partial x} + y' \frac{\partial}{\partial y}, \tag{3.5.20}$$

by using the frames

$$\boldsymbol{e}_1 = y \frac{\partial}{\partial x}, \qquad \boldsymbol{e}_2 = y \frac{\partial}{\partial y} \tag{3.5.21}$$

corresponding to θ^1 and θ^2. Thus we can write (3.5.20) as

$$x' \frac{\partial}{\partial x} + y' \frac{\partial}{\partial y} = \frac{x'}{y} \boldsymbol{e}_1 + \frac{y'}{y} \boldsymbol{e}_2. \tag{3.5.22}$$

By substituting $\xi^1 = \frac{x'}{y}, \xi^2 = \frac{y'}{y}$ into the the geodesics' differential equation

$$\begin{cases} \dfrac{d\xi^1}{ds} + \dfrac{{\omega_2}^1}{ds} \xi^2 = 0, \\[2ex] \dfrac{d\xi^2}{ds} + \dfrac{{\omega_1}^2}{ds} \xi^1 = 0, \end{cases} \tag{3.5.23}$$

we get

$$\begin{cases} \left(\dfrac{x'}{y}\right)' - \dfrac{x'}{y} \dfrac{y'}{y} = 0, \\[2ex] \left(\dfrac{y'}{y}\right)' + \dfrac{x'}{y} \dfrac{x'}{y} = 0. \end{cases} \tag{3.5.24}$$

In order to solve this differential equation we set

$$X = \frac{x'}{y}, \qquad Y = \frac{y'}{y}, \tag{3.5.25}$$

and rewrite (3.5.24) as

$$\begin{cases} X' - XY = 0, \\ Y' + XX = 0, \end{cases} \tag{3.5.26}$$

and by writing the first equation as

$$\frac{X'}{X} = Y = \frac{y'}{y} \tag{3.5.27}$$

and integrating it, we get $X = cy$, where c is a constant. Since we set the parameter s in such a way that the length of the tangent vector will be 1, we have $(\xi^1)^2+(\xi^2)^2 = 1$, i.e.,

$$X^2 + Y^2 = 1. \tag{3.5.28}$$

By substituting $X = cy$ to (3.5.28) we get

$$c^2y^2 + \left(\frac{y'}{y}\right)^2 = 1, \tag{3.5.29}$$

hence

$$ds = \frac{dy}{y\sqrt{1-c^2y^2}}. \tag{3.5.30}$$

Here, by setting

$$y = \frac{1}{c}\sin t, \tag{3.5.31}$$

we obtain

$$ds = \frac{dt}{\sin t}. \tag{3.5.32}$$

It is easy to integrate this equation, but this is not necessary. We only need the relation between ds and dt. Next,

$$y' = \frac{dy}{dt}\frac{dt}{ds} = \frac{1}{c}\cos t\,\sin t, \tag{3.5.33}$$

therefore

$$Y = \frac{y'}{y} = \cos t. \tag{3.5.34}$$

By substituting this into (3.5.28) we get

$$\frac{x'}{y} = X = \sin t. \tag{3.5.35}$$

Next, from (3.5.31), (3.5.32), and (3.5.35) we have

$$x = \int x'ds = \int x'\frac{ds}{dt}dt = \int \frac{1}{c}\sin t dt = -\frac{1}{c}\cos t + a, \tag{3.5.36}$$

where a is a constant. From (3.5.31) and (3.5.36) we get

$$x = -\frac{1}{c}\cos t + a, \qquad y = \frac{1}{c}\sin t. \tag{3.5.37}$$

By eliminating t and writing it as

$$(x-a)^2 + y^2 = \frac{1}{c^2}, \tag{3.5.38}$$

we see that the geodesic is a half circle as in Figure 3.5.6. Lastly, when going from (3.5.26) to (3.5.27), we neglected the case where $X = 0$. When $X = 0$, $x = a$ (constant) and we get a line perpendicular to the x-axis. Thus we have proved that a geodesic of the upper-half plane with the Poincaré metric is either a half circle with its diameter on the x-axis, or a line perpendicular to the x-axis. ♦

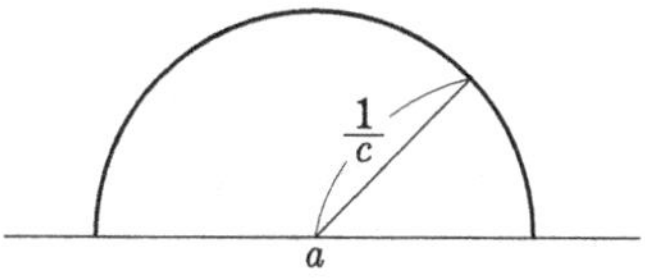

Fig. 3.5.6

Here, we would discuss a little about the relation between Euclid's axioms and geometry of the upper-half plane with the Poincaré metric.

Euclid's fifth axiom (axiom of parallel lines) says that "for any point P not in a given straight line ℓ, we can draw only one line ℓ' through P which is parallel to ℓ," but for a long time it was thought that this axiom can be proved from the other four. However from 1826 through 1829 Lobačevskiĭ (Russia), and from 1829 through 1832 Bolyai (Hungary) found that this axiom of parallel lines is independent from the other axioms, i.e., they found the existence of Non-Euclidean geometry.

By considering the above Poincaré's upper-half plane as a plane, and by calling a geodesic a line, we can draw many lines such as $\ell', \ell'', \ldots$ through a point P which are parallel to (i.e., not intersecting) the line ℓ, showing an example of geometry where the Euclid's axiom of parallel lines does not hold. This model of a Non-Euclidean geometry model by Poincaré is a very important example of a Riemannian space which is used in many different fields of mathematics.

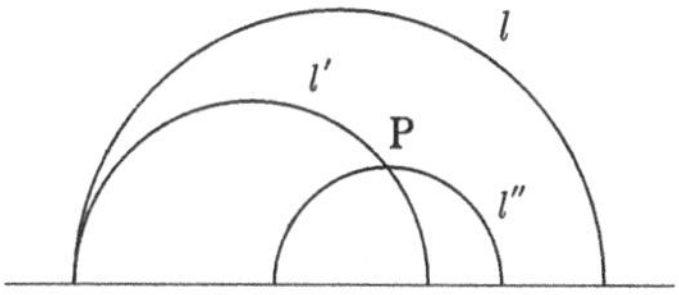

Fig. 3.5.7

Problem 3.5.1 Let $(u(t), v(t))$ be a curve on a surface with parameter t which is not necessarily the length parameter. Let $\tilde{X} = u'(t)\frac{\partial}{\partial u} + v'(t)\frac{\partial}{\partial v}$ be its tangent vector field (see (3.3.3)). Show that the geodesic equation (3.5.7) can be written as follows.

$$\frac{D\tilde{X}}{dt} + \lambda\tilde{X} = \mathbf{0},$$

where λ is a function on the curve.

Problem 3.5.2 Using the Christoffel symbols Γ^i_{jk} (see Problems 2.4.1 and 3.4.2), show that the geodesic equation (3.5.7) can be written as follows.

$$\frac{d^2u^i}{ds^2} + \sum_{j,k=1}^{2} \Gamma^i_{jk} \frac{du^j}{ds} \frac{du^k}{ds} = 0 \qquad (i = 1, 2).$$

Problem 3.5.3 Prove that every geodesic of a sphere is a great circle, and vice versa. (In Example 3.5.2 we used the argument on uniqueness of the solution of differential equations, but consider a proof avoiding this argument. By using Example 3.5.2 it is sufficient to show that every geodesic becomes a plane curve.)

3.6 Geodesics as Shortest Curves

In the previous section, we studied curves with the geodesic curvature zero, i.e., studied a geodesic as a curve where its tangent vector field is parallel along the curve itself. Here we generalized the idea where we see lines as "straight, not curved" lines on a plane, but the other important idea is that among curves connecting two points, a straight line is the shortest one. There might be readers who have already heard the expression "the shortest curve connecting two points is a geodesic," thus in this section we will make this point clear.

First, we will review the basic concepts of local maximum and minimum of a function. It is well known that if a function $L(\lambda)$ which is differentiable within interval $(-\delta, \delta)$, attains a local maximum (minimum) at the origin $\lambda = 0$, then $L'(0) = 0$. On the other hand, when $L'(0) = 0$, $\lambda = 0$ is called a critical point but a local maximum (minimum) may not be attained there, and when $L''(0) < 0$ (or > 0) a local maximum (minimum) is obtained at $\lambda = 0$. (There are more possibilities when $L''(0) = 0$ but it does not affect us now.)

From here, for simplicity we consider surfaces only in the space, and explain the case of general surfaces later. On a surface $\boldsymbol{p}(u, v)$ take two points P, Q and a curve $\boldsymbol{p}(t)$ from P to Q. Let us consider its variation, i.e., a family of curves $\boldsymbol{p}_\lambda(t)$ $(-\delta < \lambda < \delta, 0 \leq t \leq l)$ where

$$\boldsymbol{p}(t) = \boldsymbol{p}_0(t), \quad \boldsymbol{p}_\lambda(0) = \mathrm{P}, \quad \boldsymbol{p}_\lambda(l) = \mathrm{Q}. \tag{3.6.1}$$

To consider $\boldsymbol{p}_\lambda(t)$ as a function for two variables (λ, t), set

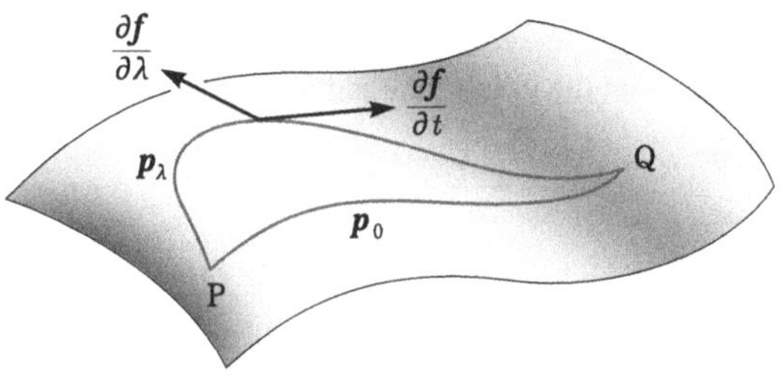

Fig. 3.6.1

$$f(\lambda, t) = \boldsymbol{p}_\lambda(t). \tag{3.6.2}$$

For a curve $\boldsymbol{p}_\lambda$ at a fixed λ, we denote by $L(\lambda)$ its length. Thus,

$$L(\lambda) = \int_0^l \left| \frac{\partial f(\lambda, t)}{\partial t} \right| dt, \tag{3.6.3}$$

where $|\ |$ denotes the length of a vector. Since the interval $(0, l)$ for t may be changed according to how parameter t is set,

$$l = L(0), \tag{3.6.4}$$

i.e., let l be the length of the curve $\boldsymbol{p}_0$. Furthermore, let parameter t of $\boldsymbol{p}_0$ represent its length, i.e., the length of $\frac{d\boldsymbol{p}_0(t)}{dt} = \frac{\partial f(0,t)}{\partial t}$ is equal to 1,

$$\left| \frac{\partial f(0, t)}{\partial t} \right| = 1. \tag{3.6.5}$$

However, generally $L(\lambda) \neq l$ for $\lambda \neq 0$, thus we cannot express the length of $\boldsymbol{p}_\lambda$ with t but it is possible to assume that t is proportional to the length of $\boldsymbol{p}_\lambda$, i.e., to assume $\left| \frac{d\boldsymbol{p}_\lambda(t)}{dt} \right|$ is a constant (which depends on λ but not on t). However, this is not important now. Next, let $f(\lambda, t)$ be a differentiable function for two variables (λ, t). (To be precise, for t it needs only to be piecewise differentiable, but we would leave that aside for now.) To prepare for calculation $L'(0)$, let us prove

$$\frac{D}{\partial \lambda} \frac{\partial f}{\partial t} = \frac{D}{\partial t} \frac{\partial f}{\partial \lambda}. \tag{3.6.6}$$

First, the covariant derivative $\frac{D}{\partial \lambda} \frac{\partial f}{\partial t}$ of the tangent vector field $\frac{\partial f}{\partial t}$ with direction λ, is the tangential component when decomposing $\frac{\partial}{\partial \lambda} \frac{\partial f}{\partial t}$ into the tangential component and the normal component (see Equation (3.4.4)), and similarly, $\frac{D}{\partial t} \frac{\partial f}{\partial \lambda}$ is the tangential component of $\frac{\partial}{\partial t} \frac{\partial f}{\partial \lambda}$. Thus, since the tangential components on both sides are equal, we obtain (3.6.6) from

$$\frac{\partial}{\partial \lambda} \frac{\partial f}{\partial t} = \frac{\partial}{\partial t} \frac{\partial f}{\partial \lambda}.$$

Express $L(\lambda)$ as

$$L(\lambda) = \int_0^l \left(\frac{\partial f(\lambda, t)}{\partial t} \cdot \frac{\partial f(\lambda, t)}{\partial t} \right)^{\frac{1}{2}} dt, \tag{3.6.7}$$

and calculate $L'(0)$ as follows.

$$
\begin{aligned}
L'(\lambda) &= \frac{d}{d\lambda}\int_0^l \left(\frac{\partial f(\lambda,t)}{\partial t}\cdot\frac{\partial f(\lambda,t)}{\partial t}\right)^{\frac{1}{2}} dt \\
&= \int_0^l \frac{\partial}{\partial\lambda}\left(\frac{\partial f(\lambda,t)}{\partial t}\cdot\frac{\partial f(\lambda,t)}{\partial t}\right)^{\frac{1}{2}} dt \\
&= \int_0^l \frac{\dfrac{\partial}{\partial\lambda}\left(\dfrac{\partial f}{\partial t}\cdot\dfrac{\partial f}{\partial t}\right)}{2\left(\dfrac{\partial f}{\partial t}\cdot\dfrac{\partial f}{\partial t}\right)^{\frac{1}{2}}} dt.
\end{aligned}
\tag{3.6.8}
$$

Then we calculate only the numerator in the integral. Using Equations (3.4.12) and (3.6.6) we obtain

$$
\begin{aligned}
&\frac{\partial}{\partial\lambda}\left(\frac{\partial f(\lambda,t)}{\partial t}\cdot\frac{\partial f(\lambda,t)}{\partial t}\right) \\
&\quad = 2\left(\frac{D}{\partial\lambda}\frac{\partial f(\lambda,t)}{\partial t}\cdot\frac{\partial f(\lambda,t)}{\partial t}\right) \\
&\quad = 2\left(\frac{D}{\partial t}\frac{\partial f(\lambda,t)}{\partial\lambda}\cdot\frac{\partial f(\lambda,t)}{\partial t}\right) \\
&\quad = 2\frac{\partial}{\partial t}\left(\frac{\partial f(\lambda,t)}{\partial\lambda}\cdot\frac{\partial f(\lambda,t)}{\partial t}\right) - 2\left(\frac{\partial f(\lambda,t)}{\partial\lambda}\cdot\frac{D}{\partial t}\frac{\partial f(\lambda,t)}{\partial t}\right).
\end{aligned}
\tag{3.6.9}
$$

Substituting this into the numerator of (3.6.8), and (3.6.5) into the denominator, we obtain

$$
\begin{aligned}
L'(0) &= \int_0^l \left\{\frac{\partial}{\partial t}\left(\frac{\partial f}{\partial\lambda}\cdot\frac{\partial f}{\partial t}\right)_{\lambda=0} - \left(\frac{\partial f}{\partial\lambda}\cdot\frac{D}{\partial t}\frac{\partial f}{\partial t}\right)_{\lambda=0}\right\} dt \\
&= \left[\left(\frac{\partial f}{\partial\lambda}\cdot\frac{\partial f}{\partial t}\right)_{\lambda=0}\right]_{t=0}^{t=l} - \int_0^l \left(\frac{\partial f}{\partial\lambda}\cdot\frac{D}{\partial t}\frac{\partial f}{\partial t}\right)_{\lambda=0} dt.
\end{aligned}
\tag{3.6.10}
$$

By (3.6.1) we have $\boldsymbol{f}(\lambda,0) = \mathrm{P}$ and $\boldsymbol{f}(\lambda,l) = \mathrm{Q}$. Hence

$$
\frac{\partial f(\lambda,l)}{\partial\lambda} = \frac{\partial f(\lambda,0)}{\partial\lambda} = \mathbf{0}. \tag{3.6.11}
$$

Therefore the first term of the right hand side of (3.6.10) vanishes and we obtain the following fundamental equation (so called **the first variation of the arc length**).

$$
L'(0) = -\int_0^l \left(\frac{\partial f}{\partial\lambda}\cdot\frac{D}{\partial t}\frac{\partial f}{\partial t}\right)_{\lambda=0} dt. \tag{3.6.12}
$$

If $\boldsymbol{p}(t) = \boldsymbol{f}(0,t)$ is a geodesic, (3.5.7) implies

$$
\frac{D}{\partial t}\frac{\partial f(0,t)}{\partial t} = \mathbf{0}. \tag{3.6.13}
$$

Thus $L'(0) = 0$.

As for the converse, $\boldsymbol{p}(t)$ is not necessarily a geodesic if $L'(0) = 0$ for a variation $\boldsymbol{p}_\lambda(t)$. However, when we fix a curve $\boldsymbol{p}(t)$, $\boldsymbol{p}(t)$ is a geodesic if $L'(0) = 0$ for all variations, that is, for all families of curves $\boldsymbol{p}_\lambda(t)$ satisfying (3.6.1).

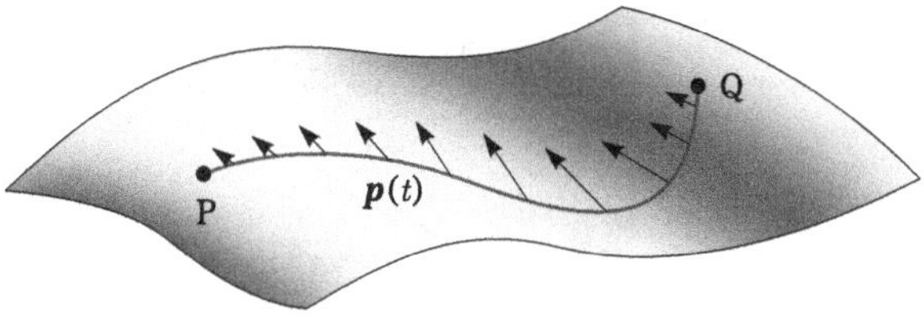

Fig. 3.6.2

In order to prove this, first we note that $\left(\frac{\partial f}{\partial \lambda}\right)_{\lambda=0}$ in (3.6.12) is a vector field defined along a curve $\boldsymbol{p}(t) = \boldsymbol{f}(0, t)$, which vanishes at $t = 0$ and $t = l$. It is easily confirmed that conversely such a vector field can be always obtained by a family of curves $\boldsymbol{p}_\lambda(t)$ as above. (It is almost obvious that we can construct such a family of curves for such a vector field in Figure 3.6.2.) Therefore we only need to prove (3.6.13) under the assumption that

$$\int_0^l \left(Y(t) \cdot \frac{D}{\partial t}\frac{\partial \boldsymbol{f}(0,t)}{\partial t}\right) dt = 0 \tag{3.6.14}$$

for any vector field $Y(t)$ defined along $\boldsymbol{p}(t)$ satisfying $Y(0) = Y(l) = \boldsymbol{0}$. Thus we choose a function $\varphi(t)$ defined along a curve $\boldsymbol{p}(t)$ satisfying $\varphi(0) = \varphi(l) = 0$ and $\varphi(t) > 0$ for $0 < t < l$ and define $Y(t)$ as

$$Y(t) = \varphi(t)\frac{D}{\partial t}\frac{\partial \boldsymbol{f}(0,t)}{\partial t}. \tag{3.6.15}$$

Then unless $\frac{D}{\partial t}\frac{\partial f(0,t)}{\partial t} = \boldsymbol{0}$, that is, unless $\boldsymbol{p}(t)$ is a geodesic, the left-hand side of (3.6.14) is positive.

In particular, if $\boldsymbol{p}(t)$ is a shortest curve joining P and Q, we have $L'(0) = 0$ for every variation $\boldsymbol{p}_\lambda(t)$ of $\boldsymbol{p}(t)$ so $\boldsymbol{p}(t)$ is a geodesic. However, a geodesic arc from P to Q is not necessarily a shortest curve as well as such as a function not necessarily taking a locally minimal value at a critical point. As an explicit example, let us consider two points P and Q on the sphere. Unless P and Q are antipodal, there is a unique great circle through P and Q. The shorter arc of the great circle is the shortest geodesic and the longer arc is also a geodesic. If we take two points P and Q on one generating line of a cylinder, the generating line gives a shortest geodesic but there are infinitely many helicoid geodesics through P and Q (Figure 3.6.3).

However, in general, a geodesic is a shortest curve in a sufficiently small domain. In order to prove it, we consider a **geodesic parallel coordinate system**, which the given Riemannian metric is written as

$$du du + G\, dv dv \quad (\text{i.e.}, E = 1, F = 0). \tag{3.6.16}$$

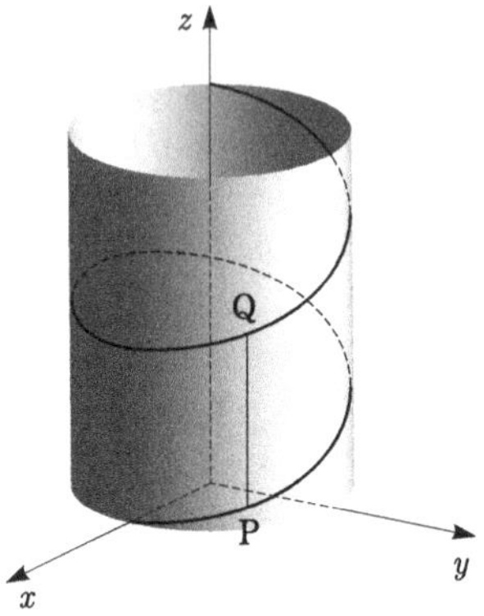

Fig. 3.6.3

In order to prove the existence of such coordinate system, at first, we take a curve C and denote its arc-length parameter by v ($|v| < \delta$). Next we draw a geodesic passing through each point of C, which orthogonally intersects with C and denote its arc-length parameter by u, needless to say let $u = 0$ on C (Figure 3.6.4). Then

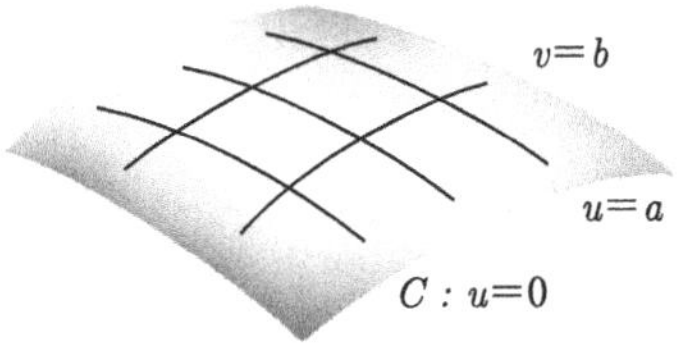

Fig. 3.6.4

it is enough to prove that $F(u, v) = 0$, i.e., the geodesics $u = a$ and $v = b$ intersect orthogonally. (Since u is the arc-length parameter, we can easily see that $E = 1$.) To prove $F(u, v) = 0$, we assume that a surface is in the space for simplicity. Differentiating $F = \boldsymbol{p}_u \cdot \boldsymbol{p}_v$ with respect to u, we obtain

$$\frac{\partial F}{\partial u} = \boldsymbol{p}_{uu} \cdot \boldsymbol{p}_v + \boldsymbol{p}_u \cdot \boldsymbol{p}_{vu} = \boldsymbol{p}_{uu} \cdot \boldsymbol{p}_v + \boldsymbol{p}_u \cdot \boldsymbol{p}_{uv}. \tag{3.6.17}$$

Since $v = b$ is a geodesic, $\boldsymbol{p}_{uu}$ is normal to the surface hence $\boldsymbol{p}_{uu} \cdot \boldsymbol{p}_v = 0$. On the other hand, since we have $0 = \boldsymbol{p}_u \cdot \boldsymbol{p}_{uv}$ by differentiating $1 = \boldsymbol{p}_u \cdot \boldsymbol{p}_u$ with respect to v, we obtain

$$\frac{\partial F}{\partial u} = 0. \tag{3.6.18}$$

However, since the curve $C : u = 0$ and the geodesic $v = b$ intersect orthogonally, we have

$$F(0, v) = 0. \tag{3.6.19}$$

These two equations imply $F(u, v) = 0$.

From the above we know not only that the Riemannian metric is written as (3.6.16) but also that

$$G(0, v) = 1. \tag{3.6.20}$$

(It is because the parameter v of C is the arc-length parameter.) Thus it may be geometrically and intuitively clear that (u, v) is a coordinate system on a range small enough. To make the argument more precise, we need to use the implicit function theorem and so on.

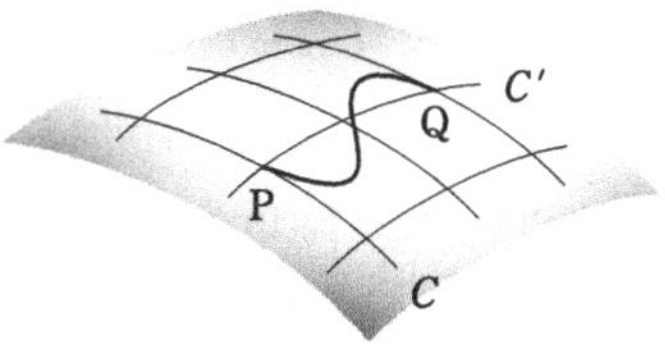

Fig. 3.6.5

Now we take a geodesic C' on a surface and fix a point P on C'. Next let C be the curve passing through P which is orthogonal to C' at P and we construct the geodesic parallel coordinate system (u, v) as stated before. (Thus C is given as $u = 0$ and C' is given as $v = 0$.) Then we join P to a point Q on C' by an arbitrary curve $(u(t), v(t))$ $(0 \leq t \leq \alpha)$ and by calculating the length we have

$$\begin{aligned}\int_0^\alpha \sqrt{\left(\frac{du}{dt}\right)^2 + G\left(\frac{dv}{dt}\right)^2}\, dt &\geq \int_0^\alpha \left|\frac{du}{dt}\right| dt \geq \left|\int_0^\alpha \frac{du}{dt}\, dt\right| \\ &= \left|\int_0^\alpha du\right| = |u(\alpha) - u(0)|.\end{aligned} \tag{3.6.21}$$

This indicates that the geodesic C' is a shortest curve joining P to Q.

In the end a few remarks are given: When we induced (3.6.12), the calculation was done under the assumption that the surface lies in the three-dimensional space. However the proof is valid when the surface lies in the higher-dimensional space if we can get the equation corresponding to (3.6.6). The equation corresponding to (3.6.6) can be obtained as follows. By fixing t and changing λ we get the vector field Λ along the curve. (This corresponds to $\frac{\partial f}{\partial \lambda}$.) Similarly we obtain the vector field T which corresponds to $\frac{\partial f}{\partial t}$.

In general, we denote by D_Λ the covariant derivative along the curve which a vector field Λ is the tangent vector field. Then we can write (3.6.6) as

$$D_\Lambda T = D_T \Lambda. \tag{3.6.22}$$

Although (3.6.22) can be shown by using the first structure equation (3.2.5), we do not discuss it further here. (We emphasize that (3.6.22) holds for arbitrary two vector fields Λ and T.)

Moreover, when we constructed the geodesic parallel coordinate system, we used the fact that the surface lies in the three-dimensional space to prove $\frac{\partial F}{\partial u} = 0$ (see (3.6.17)). In the case of general surfaces, using $U = \frac{\partial}{\partial u}, V = \frac{\partial}{\partial v}$ instead of $\boldsymbol{p}_u, \boldsymbol{p}_v$ and putting $F = \langle U, V \rangle$, we obtain

$$\frac{\partial F}{\partial u} = \frac{\partial}{\partial u}\langle U, V \rangle = \langle D_U U, V \rangle + \langle U, D_U V \rangle, \tag{3.6.23}$$

which is a substitute for (3.6.17). Since $v = b$ is a geodesic, $D_U U = 0$. On the other side, we have $D_U V = D_V U$ in a similar way to (3.6.22). By differentiating $1 = \langle U, U \rangle$ with respect to v we get $0 = \langle U, D_V U \rangle$. Then the above proof itself holds true.

Problem 3.6.1 Determine the connection form ${\omega_2}^1$ and the curvature K when a Riemannian metric is given as (3.6.16) with respect to the geodesic parallel coordinate system. (Put $g = \sqrt{G}$ for simplicity.)

Problem 3.6.2 Prove that $g_u(0, v) = 0$ if we let $C : u = 0$ be a geodesic in the construction of the geodesic parallel coordinate system. (Here $g = \sqrt{G}$, $g_u = \frac{\partial g}{\partial u}$.) Determine a Riemannian metric whose Gaussian curvature K is constant by using such coordinate systems.

Chapter 4
The Gauss-Bonnet Theorem

4.1 Integration of Exterior Differential Forms

It must have been studied in the subject of line integration in calculus that the integration of a differential 1-form

$$\varphi = f\,du + g\,dv \tag{4.1.1}$$

defined on a domain D in the (u, v)-plane along a curve

$$C : u = u(t), \qquad v = v(t) \qquad (a \leq t \leq b) \tag{4.1.2}$$

on D by

$$\int_C \varphi = \int_b^a \left(f\,\frac{du}{dt} + g\,\frac{dv}{dt} \right) dt. \tag{4.1.3}$$

For a differential 2-form

$$\psi = h\,du \wedge dv, \tag{4.1.4}$$

we define the integration of ψ on a domain A in D as

$$\int_A \psi = \iint_A h\,du\,dv. \tag{4.1.5}$$

When a differential 2-form is given as $f\,dv \wedge du$ or $f\,du \wedge dv + g\,dv \wedge du$, we apply (4.1.5) after changing it into $-f\,du \wedge dv$ or $(f - g)\,du \wedge dv$. For example,

$$\int_A f\,dv \wedge du = \int_A -f\,du \wedge dv = \iint_A -f\,du\,dv. \tag{4.1.6}$$

In the case where a transformation from a domain A in the (u, v)-plane to a domain B in the (x, y)-plane is given by functions

$$x = x(u, v), \qquad y = y(u, v), \tag{4.1.7}$$

for a differential form $kdx \wedge dy$ on B we obtain

The original version of the chapter was revised: Belated corrections have been incorporated. The correction to the chapter is available at https://doi.org/10.1007/978-981-15-1739-6_6

S. Kobayashi, *Differential Geometry of Curves and Surfaces*,
Springer Undergraduate Mathematics Series,
https://doi.org/10.1007/978-981-15-1739-6_4

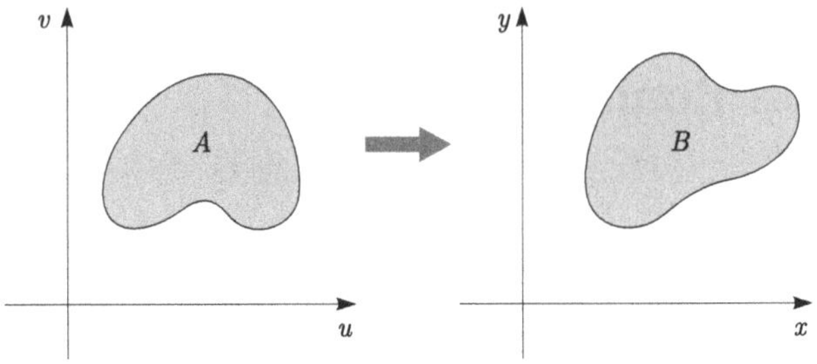

Fig. 4.1.1

$$\begin{aligned} k\,dx \wedge dy &= k\left(\frac{\partial x}{\partial u}du + \frac{\partial x}{\partial v}dv\right) \wedge \left(\frac{\partial y}{\partial u}du + \frac{\partial y}{\partial v}dv\right) \\ &= k\left(\frac{\partial x}{\partial u}\frac{\partial y}{\partial v} - \frac{\partial x}{\partial v}\frac{\partial y}{\partial u}\right) du \wedge dv \\ &= k\frac{\partial(x,y)}{\partial(u,v)} du \wedge dv, \end{aligned} \tag{4.1.8}$$

where

$$\frac{\partial(x,y)}{\partial(u,v)} = \begin{vmatrix} \frac{\partial x}{\partial u} & \frac{\partial x}{\partial v} \\ \frac{\partial y}{\partial u} & \frac{\partial y}{\partial v} \end{vmatrix}.$$

This shows that the double integral formula

$$\iint_B k(x,y)\,dx\,dy = \iint_A k(x(u,v),y(u,v))\frac{\partial(x,y)}{\partial(u,v)}\,du\,dv, \tag{4.1.9}$$

which we studied in calculus, indicates that the integral defined in (4.1.5) is invariant under the coordinate transformations. Let us note that the Jacobian $\frac{\partial(x,y)}{\partial(u,v)}$ naturally appears under the coordinate transformation when we use exterior products of differential forms.

Now, let a differential 1-form φ be written as $\varphi = df$ for a function f. Then the integration (4.1.3) becomes

$$\int_C df = f(u(b),v(b)) - f(u(a),v(a)), \tag{4.1.10}$$

which is the fundamental theorem of calculus. **Stokes' theorem**, stated in the following, is the extension to differential 2-forms. If a domain A is surrounded by a piece-wise differentiable curve C, Stokes' theorem is

$$\int_A d\varphi = \int_C \varphi \tag{4.1.11}$$

for a differential 1-form φ. We need to define the orientation of a curve C along which we integrate the line integration in the right-hand side of it (if we integrate

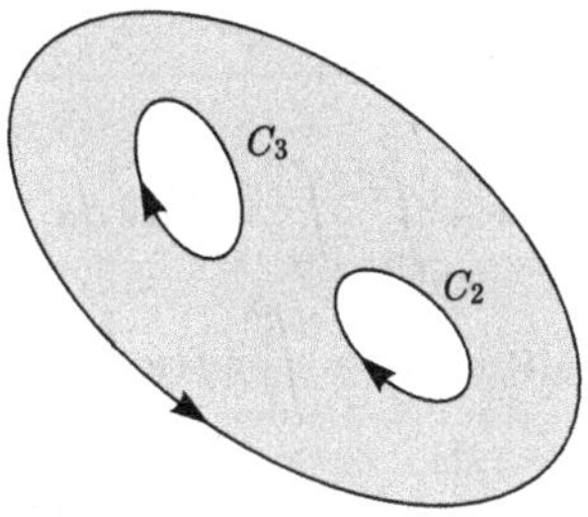

Fig. 4.1.2

along the opposite orientation, the sign of the value of integration changes). The right orientation, which is indicated in Figure 4.1.2, is the orientation for which we see the interior of A on the left when we move along C. Let ∂A denote the boundary C of the domain A with the above orientation, then equation (4.1.11) will be rewritten as

$$\int_A d\varphi = \int_{\partial A} \varphi. \tag{4.1.12}$$

Now we confirm this equation when A is a rectangle $0 \leq u \leq a$, $0 \leq v \leq b$. Set

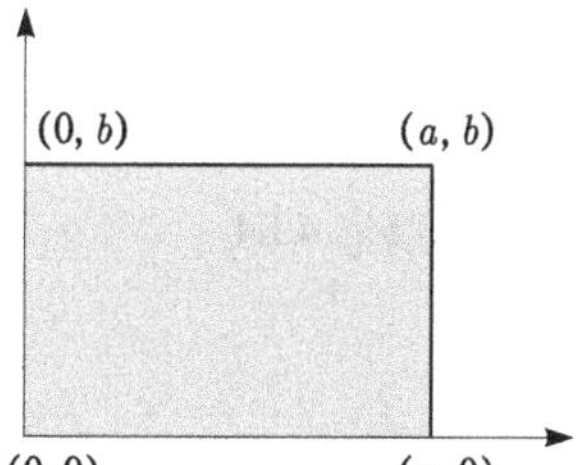

Fig. 4.1.3

$$\varphi = f\,du + g\,dv, \tag{4.1.13}$$

then

$$d\varphi = \left(\frac{\partial g}{\partial u} - \frac{\partial f}{\partial v}\right) du \wedge dv, \tag{4.1.14}$$

thus

$$\begin{aligned}\int_A d\varphi &= \int_0^b \int_0^a \left(\frac{\partial g}{\partial u} - \frac{\partial f}{\partial v}\right) du\, dv \\ &= \int_0^b \int_0^a \frac{\partial g}{\partial u}\, du\, dv - \int_0^a \int_0^b \frac{\partial f}{\partial v}\, dv\, du \\ &= \int_0^b [g(a,v) - g(0,v)]\, dv - \int_0^a [f(u,b) - f(u,0)]\, du \\ &= \int_0^a f(u,0)\, du + \int_0^b g(a,v)\, dv + \int_a^0 f(u,b)\, du + \int_b^0 g(0,v)\, dv \\ &= \int_{\partial A} \varphi.\end{aligned} \tag{4.1.15}$$

In order to prove Stokes' theorem for more general domains, we consider the case where a domain B in the (x, y)-plane differentiably and bijectively corresponds to a rectangular domain A in the (u, v)-plane. Let the correspondence F be given by

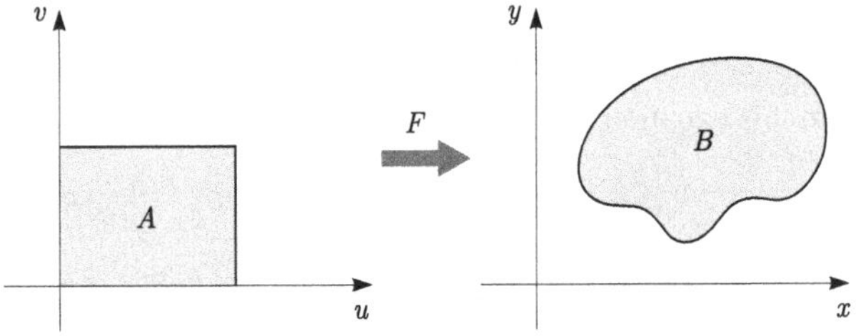

Fig. 4.1.4

functions

$$x = x(u, v), \qquad y = y(u, v). \tag{4.1.16}$$

Let

$$\psi = p(x, y)\, dx + q(x, y)\, dy \tag{4.1.17}$$

be a differential 1-form on B. The correspondence $F : A \longrightarrow B$ makes ψ change to a differential 1-form φ on A as follows.

$$\begin{aligned}\varphi = &\left[p(x(u,v), y(u,v))\frac{\partial x}{\partial u} + q(x(u,v), y(u,v))\frac{\partial y}{\partial u}\right] du \\ &+ \left[p(x(u,v), y(u,v))\frac{\partial x}{\partial v} + q(x(u,v), y(u,v))\frac{\partial y}{\partial v}\right] dv.\end{aligned} \tag{4.1.18}$$

We write it as

$$\varphi = \left(p\frac{\partial x}{\partial u} + q\frac{\partial y}{\partial u}\right) du + \left(p\frac{\partial x}{\partial v} + q\frac{\partial y}{\partial v}\right) dv \tag{4.1.19}$$

omitting variables in the description of p and q for simplicity. We also write φ as $F^*\psi$. By the definition of d (see (2.5.3)) we have

$$d\varphi = \left\{ -\frac{\partial}{\partial v}\left(p\frac{\partial x}{\partial u} + q\frac{\partial y}{\partial u}\right) + \frac{\partial}{\partial u}\left(p\frac{\partial x}{\partial u} + q\frac{\partial y}{\partial u}\right)\right\} du \wedge dv. \tag{4.1.20}$$

Calculating (4.1.20) by using $\frac{\partial p}{\partial v} = \frac{\partial p}{\partial x}\frac{\partial x}{\partial v} + \frac{\partial p}{\partial y}\frac{\partial y}{\partial v}, \cdots$ etc., and the relation $du \wedge dv = -dv \wedge du$, most terms vanish and we obtain

$$d\varphi = \left(\frac{\partial q}{\partial x} - \frac{\partial p}{\partial y}\right)\left(\frac{\partial x}{\partial u}\frac{\partial y}{\partial v} - \frac{\partial y}{\partial u}\frac{\partial x}{\partial v}\right) du \wedge dv. \tag{4.1.21}$$

On the other hand (by (2.5.3)) we have

$$d\psi = \left(\frac{\partial q}{\partial x} - \frac{\partial p}{\partial y}\right) dx \wedge dy, \tag{4.1.22}$$

and if we denote by $F^* d\psi$ the one into which ψ is transformed by F, we obtain

$$F^* d\psi = \left(\frac{\partial q}{\partial x} - \frac{\partial p}{\partial y}\right)\left(\frac{\partial x}{\partial u}\frac{\partial y}{\partial v} - \frac{\partial y}{\partial u}\frac{\partial x}{\partial v}\right) du \wedge dv \tag{4.1.23}$$

in the similar way as (4.1.8) (although the roles of (x, y) and (u, v) in (4.1.8) are exchanged). Hence $F^* d\psi$ coincides with $d\varphi$ in (4.1.21). Since $\varphi = F^*\psi$, it gives

$$F^* d\psi = d(F^*\psi). \tag{4.1.24}$$

Stokes' theorem for a domain B and a differential form ψ is proved as the following.

$$\begin{aligned} \int_B d\psi &= \int_A F^* d\psi \\ &= \int_A d(F^*\psi) \\ &= \int_{\partial A} F^*\psi = \int_{\partial B} \psi. \end{aligned} \tag{4.1.25}$$

To give a detailed explanation of it, the first equality in (4.1.25) is just (4.1.9), the second one is a result of (4.1.24), the third one is given by Stokes' theorem for a rectangular domain A, and the last one is proved as in the first equality. (Actually, the last equality is the invariance of a line integral under coordinate transformations hence the proof is much easier than the case of a double integral.) That concludes the proof of Stokes' theorem in the case where a domain B differentiably and bijectively corresponds to a rectangular domain A.

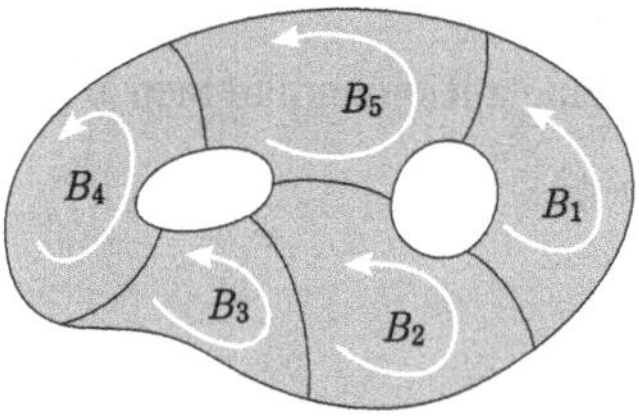

Fig. 4.1.5

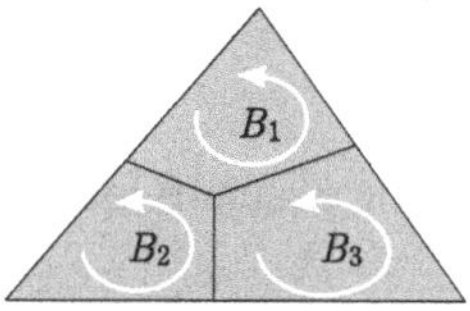

Fig. 4.1.6

As general domains, we consider the examples in Figures 4.1.5 and 4.1.6. That is, a domain B is divided into subdomains $B_1, B_2, \ldots, B_k$ and we assume that each subdomain B_i differentiably and bijectively corresponds to a rectangular domain A_i. Since we have already known that Stokes' theorem is valid for each B_i,

$$\int_B d\psi = \sum_{i=1}^{k} \int_{B_i} d\psi = \sum_{i=1}^{k} \int_{\partial B_i} \psi. \tag{4.1.26}$$

When we integrate ψ along the boundary which is the intersection of domains like B_1 and B_2, side by side with each other, we do that with the orientation of the arrows drawn in the figure thus the integration along the boundary of B_1 and that of B_2 are mutually deleted on their intersection. Hence what remains at last is only the integration along the boundary of the original B and

$$\sum_{i=1}^{k} \int_{\partial B_i} \psi = \int_{\partial B} \psi. \tag{4.1.27}$$

Stokes' theorem for B is immediately concluded by this as well as (4.1.26).

We will not deal with the problem "whether any domain B can be divided as above or not" since it is enough to only consider domains which can be divided as above for practical purposes.

Problem 4.1.1 The differential 1-form

$$\varphi = \frac{u dv - v du}{u^2 + v^2}$$

is defined on the (u, v)-plane excluding the origin.

(i) Prove $d\varphi = 0$.
(ii) Integrate φ along the unit circle with positive orientation (that is, counterclockwise orientation).
(iii) Does there exist a function f such that $df = \varphi$ on the (u, v)-plane excluding the origin?

4.2 The Gauss-Bonnet Theorem (Domains)

Assume that on a domain D in the (u, v)-plane a Riemannian metric

$$\theta^1\theta^1 + \theta^2\theta^2 \tag{4.2.1}$$

is given and the structure equation is

$$\begin{aligned} d\theta^1 &= -\omega_2{}^1 \wedge \theta^2, \\ d\theta^2 &= -\omega_1{}^2 \wedge \theta^1, \qquad \text{where} \quad \omega_1{}^2 = -\omega_2{}^1, \end{aligned} \tag{4.2.2}$$

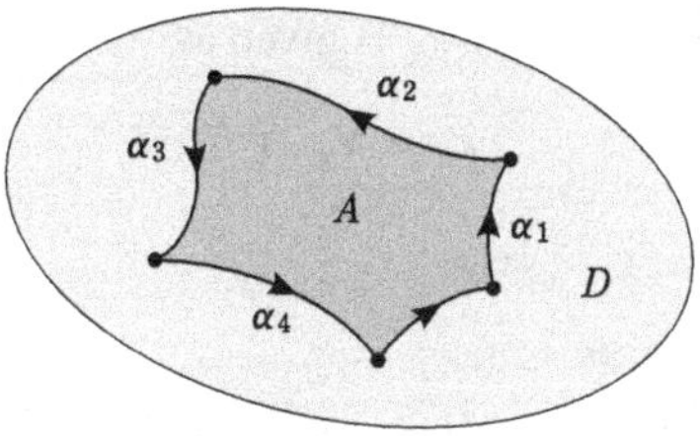

Fig. 4.2.1

$$d\omega_2{}^1 = K\,\theta^1 \wedge \theta^2 \tag{4.2.3}$$

(see Section 3.2 of Chapter 3).

We also assume that the boundary ∂A of a domain A included in D is a curve obtained by connecting (finitely) many differentiable curves $\alpha_1, \alpha_2, \ldots, \alpha_k$ like Figure 4.2.1. By Stokes's theorem in the previous section and (4.2.3) we obtain

$$\int_A K\,\theta^1 \wedge \theta^2 = \int_A d\omega_2{}^1 = \int_{\partial A} \omega_2{}^1. \tag{4.2.4}$$

Our aim is to express the line integral on the right hand side through geometric quantity. In order to simplify the notations, we fix an α_i for a while and simply denote it by $\boldsymbol{\alpha}$. We also assume a parameter s of the curve $\boldsymbol{\alpha}$ is the arc length. Thus $\boldsymbol{\alpha}'(s)$ is a unit tangent vector. Let $\boldsymbol{e}_1, \boldsymbol{e}_2$ be the orthonormal system corresponding to θ^1, θ^2 and write

$$\boldsymbol{\alpha}' = \xi^1\,\boldsymbol{e}_1 + \xi^2\,\boldsymbol{e}_2, \qquad \text{where} \quad \xi^1\,\xi^1 + \xi^2\,\xi^2 = 1. \tag{4.2.5}$$

Put

$$\boldsymbol{\nu} = -\xi^2\,\boldsymbol{e}_1 + \xi^1\,\boldsymbol{e}_2. \tag{4.2.6}$$

Then $\boldsymbol{\nu}$ is a unit vector field orthogonal to $\boldsymbol{\alpha}'$. The geodesic curvature vector (see (3.5.4))

$$\boldsymbol{k}_g = \left(\frac{d\xi^1}{ds} + \frac{\omega_2{}^1}{ds}\,\xi^2\right)\boldsymbol{e}_1 + \left(\frac{d\xi^2}{ds} + \frac{\omega_1{}^2}{ds}\,\xi^1\right)\boldsymbol{e}_2 \tag{4.2.7}$$

is a scalar multiple of $\boldsymbol{\nu}$ and we can write it as

$$\boldsymbol{k}_g = \kappa_g\,\boldsymbol{\nu} \tag{4.2.8}$$

(see (3.5.9)). Since we chose $\boldsymbol{e}_1, \boldsymbol{e}_2$ so that $\boldsymbol{e}_1 = \boldsymbol{\alpha}'$, $\boldsymbol{e}_2 = \boldsymbol{\nu}$ respectively there, let us check it directly here. In order to do this we only need to prove that $\boldsymbol{k}_g$ and $\boldsymbol{\alpha}'$ are perpendicular to each other, hence we calculate

$$\boldsymbol{k}_g \cdot \boldsymbol{\alpha}' = \frac{d\xi^1}{ds}\,\xi^1 + \frac{d\xi^2}{ds}\,\xi^2 + \left(\frac{\omega_2{}^1}{ds} + \frac{\omega_1{}^2}{ds}\right)\xi^1\xi^2, \tag{4.2.9}$$

and we obtain $\boldsymbol{k}_g \cdot \boldsymbol{\alpha}' = 0$ by $\omega_1{}^2 = -\omega_2{}^1$ and

$$0 = \frac{d}{ds}(\xi^1\,\xi^1 + \xi^2\,\xi^2) = 2\left(\frac{d\xi^1}{ds}\,\xi^1 + \frac{d\xi^2}{ds}\,\xi^2\right)$$

as well. Since the geodesic curvature κ_g is given as

$$\kappa_g = \boldsymbol{k}_g \cdot \boldsymbol{\nu} \tag{4.2.10}$$

by (4.2.8), we obtain

$$\begin{aligned} \kappa_g\, ds &= \boldsymbol{k}_g\, ds \cdot \boldsymbol{\nu} \\ &= [(d\xi^1 + {\omega_2}^1 \xi^2)\boldsymbol{e}_1 + (d\xi^2 + {\omega_1}^2 \xi^1)\boldsymbol{e}_2] \cdot [-\xi^2 \boldsymbol{e}_1 + \xi^1 \boldsymbol{e}_2] \\ &= -d\xi^1 \cdot \xi^2 - {\omega_2}^1 \xi^2\xi^2 + d\xi^2 \cdot \xi^1 + {\omega_1}^2 \xi^1\xi^1 \\ &= \xi^1\, d\xi^2 - \xi^2\, d\xi^1 - {\omega_2}^1 \end{aligned} \tag{4.2.11}$$

by using (4.2.6) and (4.2.7). Let φ denote the angle measured from $\boldsymbol{e}_1$ to $\boldsymbol{\alpha}'$. Since

$$\xi^1 = \cos\varphi, \qquad \xi^2 = \sin\varphi, \tag{4.2.12}$$

we have

$$\begin{aligned} d\xi^1 &= -\sin\varphi\, d\varphi, \\ d\xi^2 &= \cos\varphi\, d\varphi, \end{aligned} \tag{4.2.13}$$

$$\xi^1\, d\xi^2 - \xi^2\, d\xi^1 = d\varphi. \tag{4.2.14}$$

Thus we obtain

$${\omega_2}^1 = d\varphi - \kappa_g\, ds \tag{4.2.15}$$

by (4.2.11) and (4.2.14).

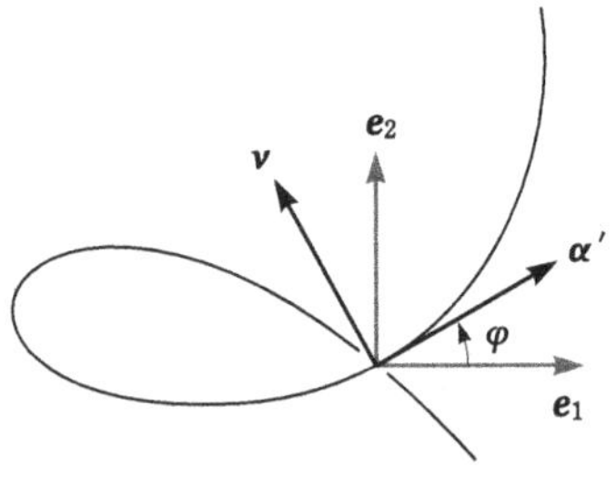

Fig. 4.2.2

Now we first prove

$$\int_\alpha d\varphi = 2\pi \tag{4.2.16}$$

in the case where ∂A is differentiable, that is, ∂A is a smooth curve $\alpha(t)\,(0 \le t \le a)$ and it closes smoothly at both the beginning point $\alpha(0)$ and the end point $\alpha(a)$. When we take an angle $\varphi(t)$ on the unit circle, $\varphi(t)$ moves from $\varphi(0)$ to $\varphi(a)$ on the unit circle as t moves from 0 to a. Since $\alpha(t)$ is smooth at $\alpha(0) = \alpha(a)$, we have $\varphi(0) = \varphi(a)$. Hence $\varphi(t)$ rotates several times on the unit circle (although it may go back and forth) and returns to the original position. If we denote this number of

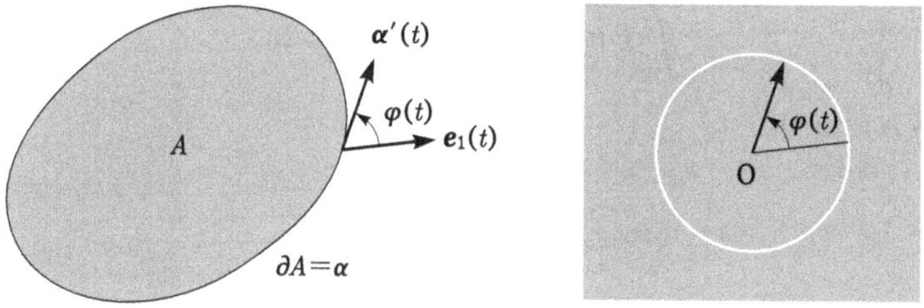

Fig. 4.2.3

rotation by n, the integral in the left-hand side of (4.2.16) should coincide with $2n\pi$. (This is clear if we consider the meaning of the integral $\int_{\partial A} d\varphi$.) Hence we need only to show $n = 1$ to prove (4.2.16).

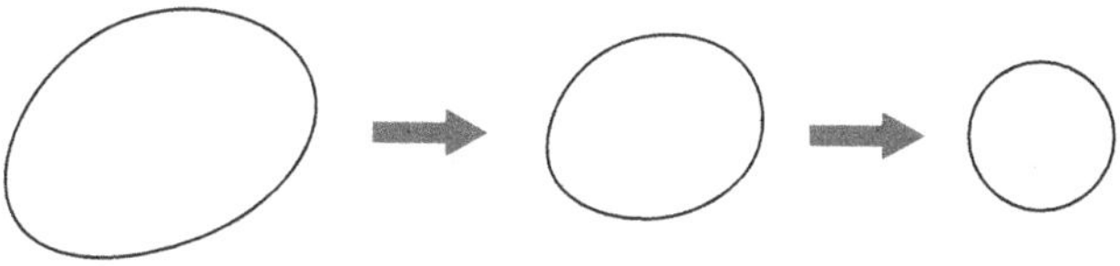

Fig. 4.2.4

In order to do this, we gradually deform α to a small circle in the plane. When we deform α continuously, the integral in the left-hand side of (4.2.16) changes continuously. On the other hand, the value of the integral is $2n\pi$ which cannot change gradually since n is an integer and it is a constant number $2n\pi$. Hence the integral in the left-hand side of (4.2.16) is $2n\pi$ if α is a small circle in the plane, so it is enough to consider the case where α is a small circle in the plane to show $n = 1$. We clearly have $n = 1$ since $\boldsymbol{e}_1$ hardly changes its direction on a small circle whereas the tangent vector of the circle rotates only once.

There may be readers who are worried that $\varphi(t)$ is an angle measured with the Riemannian metric given on a domain D but not a Euclidean angle on a domain D in the Euclidean plane. In order to justify it, we gradually deform the given metric $ds^2 = \theta^1\,\theta^1 + \theta^2\,\theta^2$ to the Euclidean metric $d\sigma^2 = dudu + dvdv$. That is, we give

$$ds_\tau{}^2 = (1-\tau)ds^2 + \tau d\sigma^2 \qquad (0 \le \tau \le 1) \tag{4.2.17}$$

and we let the metric ds^2 change to the metric $d\sigma^2$ when the parameter τ changes from 0 to 1. If we measure φ in the left-hand side in (4.2.16) by the metric $ds_\tau{}^2$ and change τ, the value $2n\pi$ of the integral must change gradually. On the other hand, since n is an integer, the value should be a constant number, as in the above argument where we gradually changed α. Hence we may consider the integral (4.2.16) with respect to a ordinary Euclidean angle φ.

By (4.2.4), (4.2.15) and (4.2.16) we obtain the formula

$$\int_A K\,\theta^1 \wedge \theta^2 + \int_{\partial A} \kappa_g\, ds = 2\pi \tag{4.2.18}$$

when ∂A is smooth.

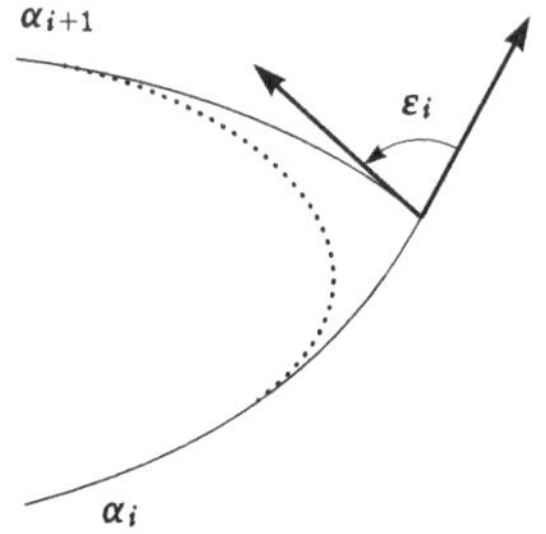

Fig. 4.2.5

Next we consider the case where ∂A is parted in n smooth curves $\alpha_1, \alpha_2, \ldots, \alpha_n$. Let ε_i denote an angle (i.e., an exterior angle) formed when moving from α_i to α_{i+1} (the exterior angle formed when moving from α_n to α_1 is ε_n.) We denote by B a domain obtained by smoothing each corner like the dotted line in Figure 4.2.5. We only change the curve near a corner and keep B as an approximation of A. Since ∂B is smooth, it is the case we have already considered and we obtain

$$\int_{\partial B} d\varphi = 2\pi \tag{4.2.19}$$

by (4.2.16). Now we divide ∂B into two parts, the original part of ∂A and the dotted curves. Let β' denote $\partial B \cap \partial A$, β'' denote the dotted curves and we rewrite (4.2.19) as

$$\int_{\partial B} d\varphi = \int_{\beta'} d\varphi + \int_{\beta''} d\varphi = 2\pi. \tag{4.2.20}$$

By making β'' smaller and smaller, β' approaches ∂A and the integral $\int_{\beta'} d\varphi$ approaches $\sum_{i=1}^n \int_{\alpha_i} d\varphi$. We write this as

$$\lim \int_{\beta'} d\varphi = \sum_{i=1}^{n} \int_{\alpha_i} d\varphi = \int_{\partial A} d\varphi. \tag{4.2.21}$$

Considering what happens to $\lim \int_{\beta'} d\varphi$ at this moment, we obtain

$$\lim \int_{\beta''} d\varphi = \sum_{i=1}^{n} \varepsilon_i, \tag{4.2.22}$$

since the integral is the sum of the quantity of change of φ on each dotted curve. By (4.2.20), (4.2.21) and (4.2.22) we obtain

$$2\pi = \lim \int_{\partial B} d\varphi = \int_{\partial A} d\varphi + \sum_{i=1}^{n} \varepsilon_i. \tag{4.2.23}$$

(The meaning of this equation is that the total continuous variation of φ is obtained as $\int_{\partial A} d\varphi$ and the total quantity of the sudden change at each corner is written as $\sum_{i=1}^{n} \varepsilon_i$ and their sum comes to be one rotation, that is, 2π.) Hence we obtain the **Gauss-Bonnet theorem**

$$\int_A K\,\theta^1 \wedge \theta^2 + \int_{\partial A} \kappa_g\, ds = 2\pi - \sum_{i=1}^{n} \varepsilon_i \tag{4.2.24}$$

by (4.2.4), (4.2.15) and (4.2.23). It is a generalization of (4.2.18). In the above proof we made ∂A smaller and smaller and finally we changed it to a small circle. Then we used the fact that the direction of the vector field $\boldsymbol{e}_1$ is almost constant on the circle. For example, when there are some holes or a slit like Figures 4.2.6 or 4.2.7, ∂A cannot approach a small circle since it stops at a hole or a slit when we make ∂A smaller. In the case where only a point P is removed like Figure 4.2.8, ∂A possibly approaches a small circle whose center is P but we cannot say the direction of the vector field $\boldsymbol{e}_1$ is almost constant on the small circle since $\boldsymbol{e}_1$ is not defined at P. It may be possible that $\boldsymbol{e}_1$ has a radial direction from P for instance. Then $\boldsymbol{e}_1$ is continuously defined outside of P but we cannot define $\boldsymbol{e}_1$ at P so that it is continuous at P too. As going around the small circle centering P once, far from being constant, $\boldsymbol{e}_1$ rotates with 2π. Hence the Gauss-Bonnet theorem cannot hold for domains as above.

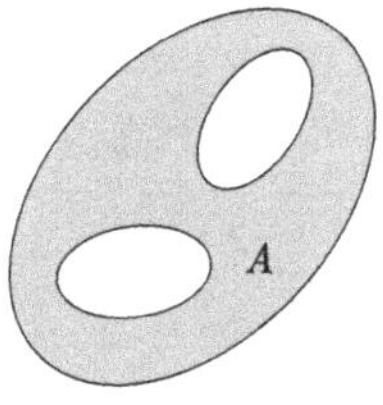

Fig. 4.2.6

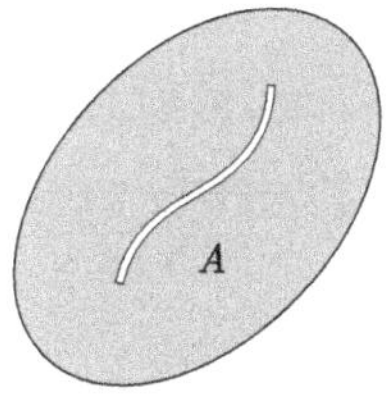

Fig. 4.2.7

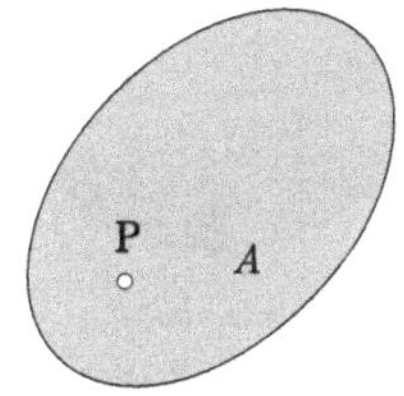

Fig. 4.2.8

A domain A for which the theorem holds is called "simply connected." That is, it is needed that ∂A can be gradually deformed through the inside of A to a small circle centering at point P in A. As noted above, it is important that P is included in A.

We apply the Gauss-Bonnet theorem (4.2.24) to the case where ∂A is a triangle (see Figure 4.2.9). Let $\iota_1, \iota_2, \iota_3$ denote interior angles at the vertices. Substituting $\varepsilon_k = \pi - \iota_k$ $(k = 1, 2, 3)$ for (4.2.24), we obtain

$$\int_A K\,\theta^1 \wedge \theta^2 + \int_{\partial A} \kappa_g\, ds = \iota_1 + \iota_2 + \iota_3 - \pi. \tag{4.2.25}$$

In particular, if the three sides $\alpha_1, \alpha_2, \alpha_3$ are geodesics, we obtain

$$\int_A K\,\theta^1 \wedge \theta^2 = \iota_1 + \iota_2 + \iota_3 - \pi. \tag{4.2.26}$$

In the case of Euclidean geometry we have $K \equiv 0$ hence (4.2.26) means that the sum of the interior angles of a usual triangle is π.

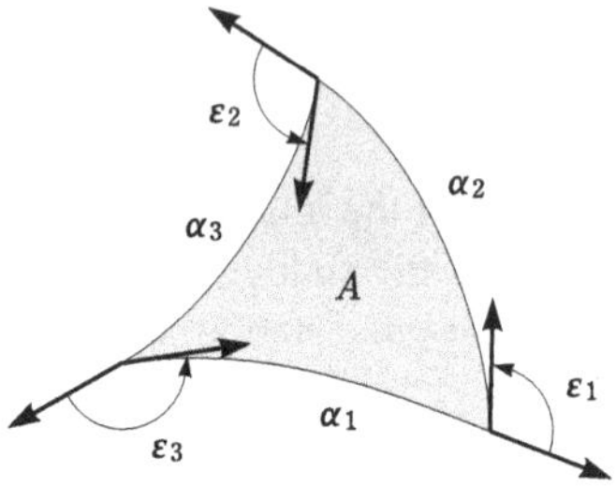

Fig. 4.2.9

If we consider a triangle where each of the sides is an arc of a great circle on a sphere, K is a positive constant number and the sum of the interior angles of the triangle is greater than π. This is a well-known fact in spherical trigonometry. Similarly, on the upper half plane endowed with Poincaré metric, the sum of the interior angles of a triangle where each of the sides is a half circle with its center on the u-axis is less than π (Figure 4.2.11). (Due to the form of the Poincaré metric, we note that the angles are the same if we measure them by the Euclidean metric.)

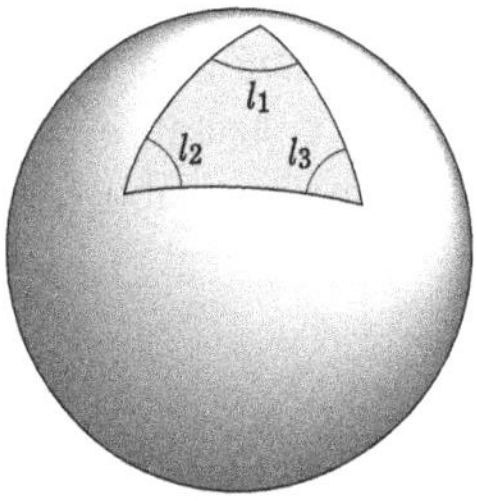

$l_1+l_2+l_3>\pi$

Fig. 4.2.10

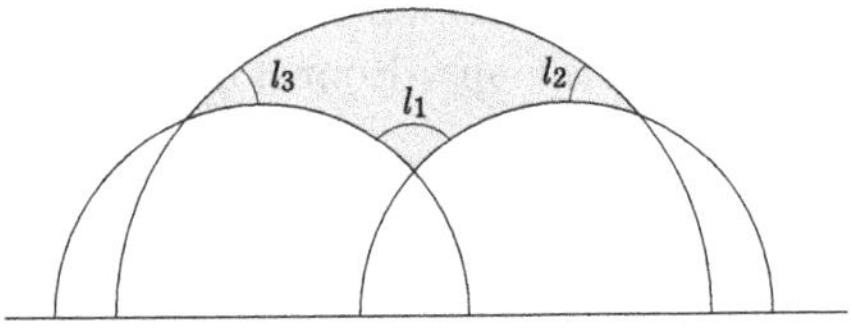

$l_1+l_2+l_3<\pi$

Fig. 4.2.11

Problem 4.2.1 We assume that on a rectangular domain in the (u,v)-plane, a Riemannian metric is given and its Gaussian curvature K is negative or zero. Prove that two geodesics never intersect at two points P, Q like Figure 4.2.12.

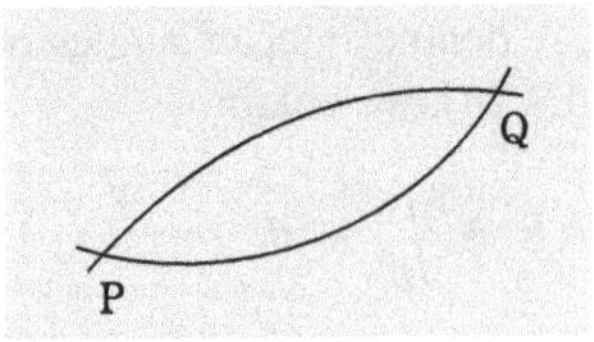

Fig. 4.2.12

Problem 4.2.2 As in Example 3.4.1 of Chapter 3 we define parameters u, v on a sphere with radius a. Calculate the geodesic curvature κ_g of a small circle obtained by fixing u, calculate the area above this small circle on the sphere, and verify the Gauss-Bonnet theorem in this case directly.

Problem 4.2.3 Prove that

$$\int_A K\theta^1 \wedge \theta^2 + \int_{\partial A} \kappa_g ds = \iota_1 + \iota_2 + \iota_3 + \iota_4 - 2\pi$$

when A is a quadrilateral and the interior angles at P_1, P_2, P_3, P_4 are $\iota_1, \iota_2, \iota_3, \iota_4$ respectively.

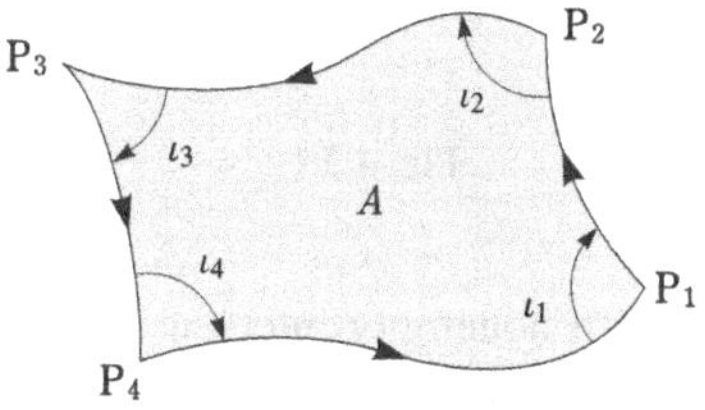

Fig. 4.2.13

4.3 The Gauss-Bonnet Theorem (Closed Surfaces)

In the previous section we proved the Gauss-Bonnet theorem for a domain in plane endowed with a Riemannian metric. Here we aim to apply it to "a closed surface endowed with orientation" and to describe "the Euler characteristic" of the surface as an integral of the curvature.

First, we try to consider the case of a surface in the space to explain what "an oriented closed surface" is. A surface S in the space is, as it is explained in Section 2.1 of Chapter 2, a union of surface pieces U_α. Here each U_α bijectively corresponds to a domain D_α in the (u, v)-plane as follows. Let $f_\alpha, g_\alpha, h_\alpha$ be functions on D and suitable times differentiable (at least three times is enough in the following). Moreover,

$$x = f_\alpha(u, v), \qquad y = g_\alpha(u, v), \qquad z = h_\alpha(u, v) \tag{4.3.1}$$

gives the correspondence between U_α and D_α and the rank of the matrix

$$\begin{bmatrix} \dfrac{\partial x}{\partial u} & \dfrac{\partial x}{\partial v} \\[2ex] \dfrac{\partial y}{\partial u} & \dfrac{\partial y}{\partial v} \\[2ex] \dfrac{\partial z}{\partial u} & \dfrac{\partial z}{\partial v} \end{bmatrix} \tag{4.3.2}$$

is two. (This was the condition for the existence of the tangent plane.) Now we

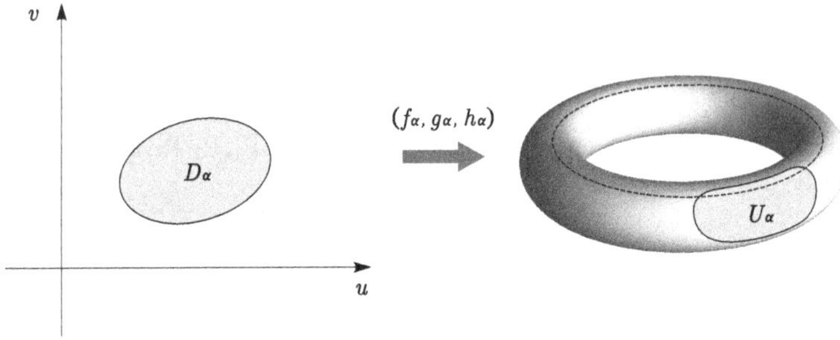

Fig. 4.3.1

assume that U_α and U_β have a nonempty intersection. Let $D_{\alpha\beta}$ denote the corresponding part in D_α to $U_\alpha \cap U_\beta$ and $D_{\beta\alpha}$ denote that in D_β. Then there is a bijective correspondence between $U_{\alpha\beta}$ and $U_{\beta\alpha}$ which is several times differentiable. In order to distinguish planes including D_α and D_β, we write the coordinate (u, v) of each plane as $(u_\alpha, v_\alpha), (u_\beta, v_\beta)$. Then the correspondence between $D_{\alpha\beta}$ and $D_{\beta\alpha}$ is written as

$$u_\beta = u_\beta(u_\alpha, v_\alpha), \qquad v_\beta = v_\beta(u_\alpha, v_\alpha). \tag{4.3.3}$$

The fact that the rank of the matrix (4.3.2) equals 2 and the formula of differentiation

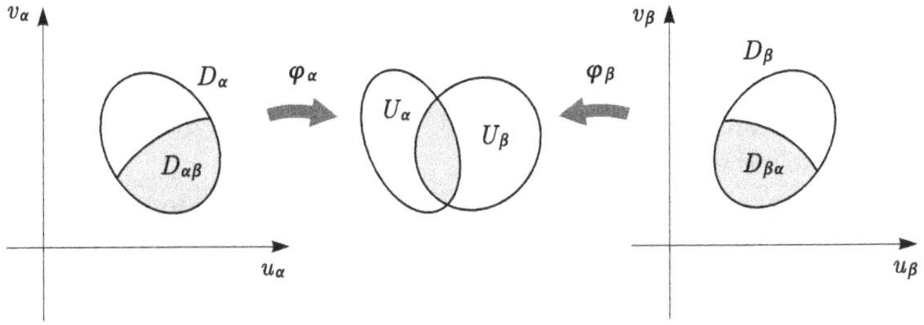

Fig. 4.3.2

$$\begin{bmatrix} \dfrac{\partial x}{\partial u_\alpha} & \dfrac{\partial x}{\partial v_\alpha} \\ \dfrac{\partial y}{\partial u_\alpha} & \dfrac{\partial y}{\partial v_\alpha} \\ \dfrac{\partial z}{\partial u_\alpha} & \dfrac{\partial z}{\partial v_\alpha} \end{bmatrix} = \begin{bmatrix} \dfrac{\partial x}{\partial u_\beta} & \dfrac{\partial x}{\partial v_\beta} \\ \dfrac{\partial y}{\partial u_\beta} & \dfrac{\partial y}{\partial v_\beta} \\ \dfrac{\partial z}{\partial u_\beta} & \dfrac{\partial z}{\partial v_\beta} \end{bmatrix} \begin{bmatrix} \dfrac{\partial u_\beta}{\partial u_\alpha} & \dfrac{\partial u_\beta}{\partial v_\alpha} \\ \dfrac{\partial v_\beta}{\partial u_\alpha} & \dfrac{\partial v_\beta}{\partial v_\alpha} \end{bmatrix} \tag{4.3.4}$$

concludes that the Jacobian of the transformation (4.3.3)

$$\frac{\partial(u_\beta, v_\beta)}{\partial(u_\alpha, v_\alpha)} = \begin{bmatrix} \dfrac{\partial u_\beta}{\partial u_\alpha} & \dfrac{\partial u_\beta}{\partial v_\alpha} \\[2ex] \dfrac{\partial v_\beta}{\partial u_\alpha} & \dfrac{\partial v_\beta}{\partial v_\alpha} \end{bmatrix} \tag{4.3.5}$$

has rank 2, that is, its determinant is not zero.

In the above we refer to properties of a surface in the space. By omitting facts relating with space coordinate (x, y, z), we obtain a definition of a surface in general. Since we forward our argument without assuming that readers have already learned the notion of topology, there are some incorrect points in the following definition of a surface but it is enough to understand it intuitively. A surface S satisfies the following:

(a) It is covered by subsets U_α, that is, $S = \bigcup_\alpha U_\alpha$.
(b) Each U_α has a bijective correspondence to a domain D_α in the (u_α, v_α)-plane. We denote the correspondence by $\varphi_\alpha : D_\alpha \to U_\alpha$.
(c) Assume that U_α and U_β have nonempty intersection. Putting $D_{\alpha\beta} = \varphi_\alpha^{-1}(U_\alpha \cap U_\beta)$ and $D_{\beta\alpha} = \varphi_\beta^{-1}(U_\alpha \cap U_\beta)$, we write the bijective correspondence $\varphi_\beta^{-1} \circ \varphi_\alpha : D_{\alpha\beta} \to D_{\beta\alpha}$ as (4.3.3). Then the rank of the Jacobian (4.3.5) is 2.

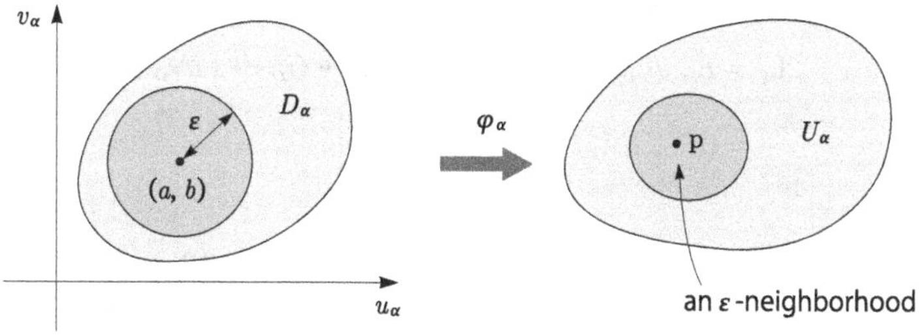

Fig. 4.3.3

In order to state another condition for a surface S, we define an ε-neighborhood of $p \in U_\alpha$ in U_α (Figure 4.3.3). It is the subset in U_α corresponding to a subset in D_α which is determined by

$$(u_\alpha - a)^2 + (v_\alpha - b)^2 < \varepsilon^2, \tag{4.3.6}$$

where (a, b) in D_α is the point corresponding to p, under φ_α. The last condition is:

(d) if $p, q \in S, p \neq q$, $p \in U_\alpha$ and $q \in U_\beta$, the ε-neighborhood of p and the ε-neighborhood of q have an empty intersection for sufficient small $\varepsilon > 0$.

From now on, we assume that a surface always satisfies the above conditions (a), (b), (c) and (d). In order to understand the condition (d), it is better to consider an example which does not satisfy (d). To construct such S, let D_1 be a circular domain

$$(u_1)^2 + (v_1)^2 < 1$$

in the (u_1, v_1)-plane. Similarly let D_2 denote a circular domain

$$(u_2)^2 + (v_2)^2 < 1$$

in the (u_2, v_2)-plane. Let D_{12} denote D_1 excluding the origin and D_{21} denote D_2 excluding the origin and we give a bijective correspondence between D_{12} and D_{21} by $u_2 = u_1$ and $v_2 = v_1$.

In other words, S is obtained by putting D_2 on top of D_1 and identifying coincidental points except for the origin. Let p be the origin of D_1 and q be that of D_2, the condition (d) is not satisfied. Hence we do not consider such an example as a surface.

Although we have been considering a Riemannian metric defined on the (u, v)-plane, we consider a surface in the space again in order to explain a Riemannian metric on a surface in general. When a surface is given by (4.3.1) (but (u, v) is denoted by (u_α, v_α) and we use a vector notation as $\boldsymbol{p} = (x, y, z)$), by substituting (4.3.8) into (4.3.7)

$$d\boldsymbol{p} \cdot d\boldsymbol{p} = dx\,dx + dy\,dy + dz\,dz, \tag{4.3.7}$$

$$d\boldsymbol{p} = \frac{\partial \boldsymbol{p}}{\partial u_\alpha} du_\alpha + \frac{\partial \boldsymbol{p}}{\partial v_\alpha} dv_\alpha, \tag{4.3.8}$$

we obtain the first fundamental formula

$$\begin{aligned} \mathrm{I}_\alpha &= E_\alpha\, du_\alpha\, du_\alpha + 2F_\alpha\, du_\alpha\, dv_\alpha + G_\alpha\, dv_\alpha\, dv_\alpha \\ &= (du_\alpha, dv_\alpha) \begin{bmatrix} E_\alpha & F_\alpha \\ F_\alpha & G_\alpha \end{bmatrix} \begin{bmatrix} du_\alpha \\ dv_\alpha \end{bmatrix}, \end{aligned} \tag{4.3.9}$$

where

$$E_\alpha = \frac{\partial \boldsymbol{p}}{\partial u_\alpha} \cdot \frac{\partial \boldsymbol{p}}{\partial u_\alpha}, \quad F_\alpha = \frac{\partial \boldsymbol{p}}{\partial u_\alpha} \cdot \frac{\partial \boldsymbol{p}}{\partial v_\alpha}, \quad G_\alpha = \frac{\partial \boldsymbol{p}}{\partial v_\alpha} \cdot \frac{\partial \boldsymbol{p}}{\partial v_\alpha}.$$

If U_α and U_β have a nonempty intersection, I_α and I_β should coincide on it since both are obtained from $d\boldsymbol{p} \cdot d\boldsymbol{p}$. Substituting

$$(du_\beta, dv_\beta) = (du_\alpha, dv_\alpha) \begin{bmatrix} \dfrac{\partial u_\beta}{\partial u_\alpha} & \dfrac{\partial v_\beta}{\partial u_\alpha} \\ \dfrac{\partial u_\beta}{\partial v_\alpha} & \dfrac{\partial v_\beta}{\partial v_\alpha} \end{bmatrix} \tag{4.3.10}$$

to I_β in $\mathrm{I}_\alpha = \mathrm{I}_\beta$ and comparing it with I_α, we obtain the relation

$$\begin{bmatrix} \dfrac{\partial u_\beta}{\partial u_\alpha} & \dfrac{\partial v_\beta}{\partial u_\alpha} \\ \dfrac{\partial u_\beta}{\partial v_\alpha} & \dfrac{\partial v_\beta}{\partial v_\alpha} \end{bmatrix} \begin{bmatrix} E_\beta & F_\beta \\ F_\beta & G_\beta \end{bmatrix} \begin{bmatrix} \dfrac{\partial u_\beta}{\partial u_\alpha} & \dfrac{\partial u_\beta}{\partial v_\alpha} \\ \dfrac{\partial v_\beta}{\partial u_\alpha} & \dfrac{\partial v_\beta}{\partial v_\alpha} \end{bmatrix} = \begin{bmatrix} E_\alpha & F_\alpha \\ F_\alpha & G_\alpha \end{bmatrix}. \tag{4.3.11}$$

A Riemannian metric on a general surface S which is not necessarily a surface in the space can be obtained by deleting a part of $\boldsymbol{p} = (x, y, z)$ from the property of the first fundamental form stated before. That is, when $S = \bigcup_\alpha U_\alpha$ holds and U_α corresponds bijectively to a domain D_α in the (u_α, v_α)-plane, we define a Riemannian metric on S by giving a Riemannian metric (4.3.9) on each D_α which satisfies the relation (4.3.11) when U_α and U_β have a nonempty intersection.

In other words, (4.3.11) means that if we measure the length of a curve on $U_\alpha \cap U_\beta$ with Riemannian metrics on both D_α and D_β, they give the same length.

Next, we explain a **triangulation** of a surface. At first, a triangle T in D_α consists of

(a) three points in D_α (which are called vertices of T),
(b) three smooth curves which join each vertex and do not intersect except for vertices, (which are called edges),
(c) the domain surrounded by three edges, (which is called a face).

Under the bijective correspondence between D_α and U_α, we call a figure corresponding to a triangle in D_α as a triangle in U_α. Here we do not consider such a large triangle which is not contained in one U_α (see Figure 4.3.4). A triangulation

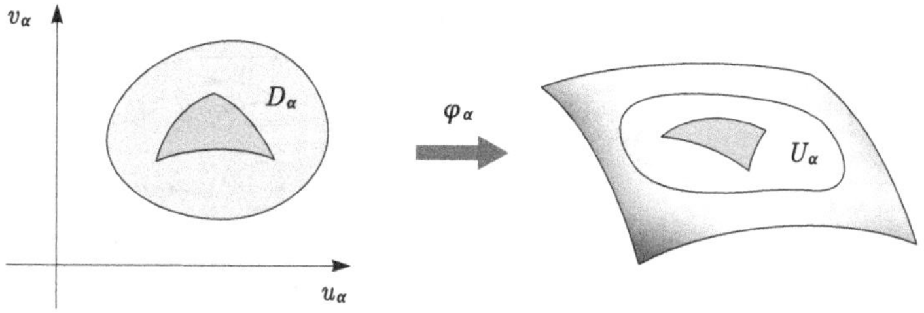

Fig. 4.3.4

of a surface S is to cover S with such triangles (that is, to express S as a union of triangles). Moreover, we assume that

(d) if a point P in S belongs to a triangle T and it does not lie on an edge, T is a unique triangle which contains P,
(e) if P lies on an edge of a triangle T and it is not a vertex, there is only one triangle T' which contains an edge that P belongs to and P is contained in an interior of $T \cup T'$,
(f) if P is a vertex of a triangle T, there are finite triangles $T = T_1, T_2, \ldots, T_k$ where T_i and T_j share only one edge (we consider $T_{k+1} = T_1$) and P is contained in an interior of $T_1 \cup \cdots \cup T_k$.

Figure 4.3.5 and 4.3.6 do not satisfy the conditions above.

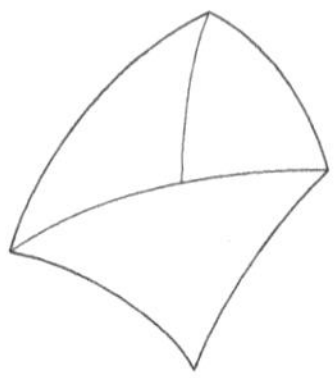

Fig. 4.3.5

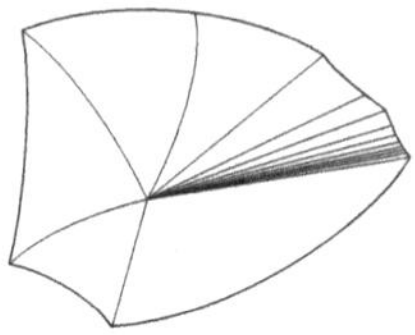

Fig. 4.3.6

It is intuitively clear that any surface has a triangulation. A strict proof of this involves complicated discussion, thus for now we shall proceed assuming that triangulation is possible.

We give an orientation on each edge of a triangle in a domain D_α so that when we move according to the orientation, we see the interior of the triangle on our left-hand side. By using the bijective correspondence between D_α and U_α we give an orientation on each edge of a triangle in U_α. Since each edge of a surface S with triangulation belongs exactly two triangles, it has two orientations. If these orientations are opposite to each other, we say that S is **oriented**. There are surfaces which cannot be oriented but we will refer to them later.

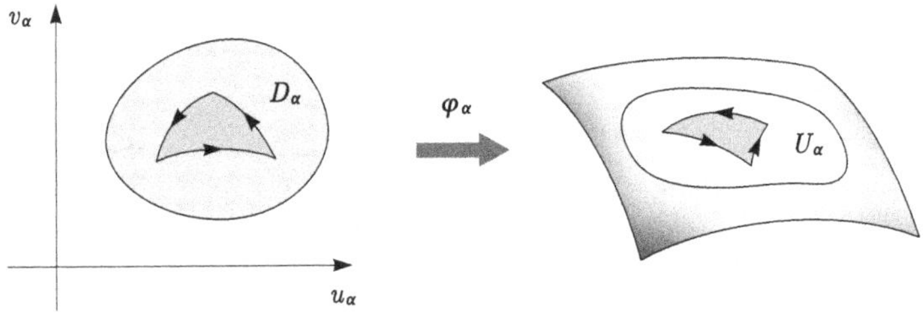

Fig. 4.3.7

We call a surface divided into finite triangles a **closed surface**. This is not the usual definition but they are the same (usually, a closed surface means a compact surface without boundary but here we avoid words from topology like "compact").

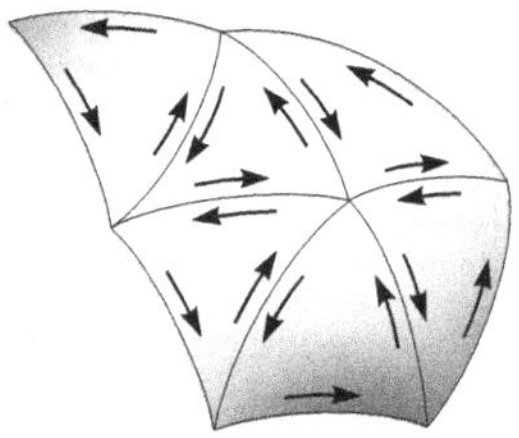

Fig. 4.3.8

Now we are ready to prove the Gauss-Bonnet theorem for an oriented closed surface. We divide S into finite triangles $T_1, T_2, \ldots, T_f$ as above and apply the Gauss-Bonnet theorem in the previous section to each T_λ (see (4.2.25)). Then we obtain

$$\int_{T_\lambda} K\,\theta^1 \wedge \theta^2 + \int_{\partial T_\lambda} \kappa_g\,ds = \iota_{\lambda 1} + \iota_{\lambda 2} + \iota_{\lambda 3} - \pi. \tag{4.3.12}$$

Here $\iota_{\lambda 1}, \iota_{\lambda 2}$ and $\iota_{\lambda 3}$ denote interior angles of the triangle T_λ. Summing up the equation from $\lambda = 1$ to $\lambda = f$, we obtain

$$\int_S K\,\theta^1 \wedge \theta^2 + \sum_{\lambda=1}^{f} \int_{\partial T_\lambda} \kappa_g\,ds = \sum_{\lambda=1}^{f} (\iota_{\lambda 1} + \iota_{\lambda 2} + \iota_{\lambda 3}) - f\pi. \tag{4.3.13}$$

In triangulation each of the edges are common to the edges of the two triangles whose orientations induced from these triangles are opposite to each other. Hence if we integrate κ_g along ∂T_λ and adding from $\lambda = 1$ to $\lambda = f$, the integration along each edge arises twice with opposite orientation and so it vanishes. Therefore we have

$$\sum_{\lambda=1}^{f} \int_{\partial T_\lambda} \kappa_g\,ds = 0. \tag{4.3.14}$$

Next, when we calculate $\sum(\iota_{\lambda 1} + \iota_{\lambda 2} + \iota_{\lambda 3})$, we get as many 2π's as the number of vertexes if we gather the angles at each vertex. Denoting the number of vertexes by v,

$$\sum_{\lambda=1}^{f} (\iota_{\lambda 1} + \iota_{\lambda 2} + \iota_{\lambda 3}) = 2v\pi. \tag{4.3.15}$$

Hence (4.3.13) comes to be the following:

$$\int_S K\,\theta^1 \wedge \theta^2 = (2v - f)\pi. \tag{4.3.16}$$

Since each triangle T_λ has three edges and each edge belongs to two triangles, the relation between the number of triangles f and that of edges e is given by

$$3f = 2e. \tag{4.3.17}$$

Hence, according to the deformation $2v - f = 2v - 3f + 2f = 2(v - e + f)$, we can write (4.3.16) as

$$\int_S K\,\theta^1 \wedge \theta^2 = 2\pi(v - e + f). \tag{4.3.18}$$

We call $v - e + f$ (the number of vertexes $-$ the number of edges $+$ the number of faces) the **Euler characteristic** of S usually writing it as $\chi(S)$. Thus we can also write (4.3.18) as

$$\int_S K\,\theta^1 \wedge \theta^2 = 2\pi\chi(S). \tag{4.3.19}$$

This is the **Gauss-Bonnet theorem** for an oriented closed surface. It indicates that the curvature K depends on the given Riemannian metric of S but its integral, the left-hand side of (4.3.19), equals to $2\pi\chi(S)$ which is the number independent of the Riemannian metric. On the other hand, $\chi(S) = v - e + f$ seems to depend on triangulations at a glance but we can clearly see from the left-hand side of (4.3.19) that $\chi(S)$ does not depend on triangulations. Now we have justified ourselves to say that $\chi(S)$ is the Euler characteristic of S indicating that it is independent from triangulations.

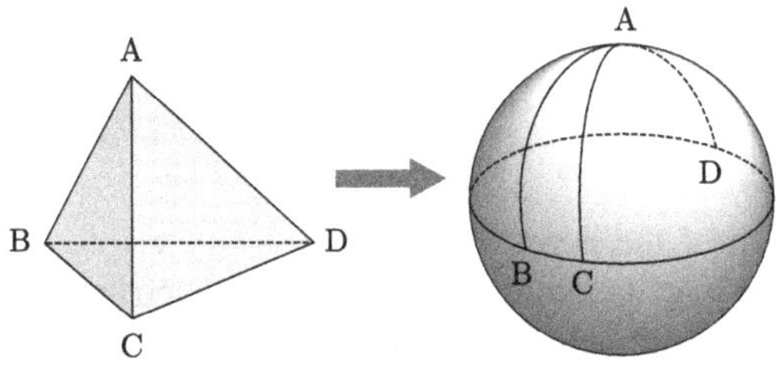

Fig. 4.3.9

Next, we simply calculate the Euler characteristic. First, if we consider the sphere as an expansion of a tetrahedron like Figure 4.3.9, we get a triangulation corresponding to the triangulation of a tetrahedron, naturally. It is easy to calculate $v - e + f$ for a tetrahedron, which is $4 - 6 + 4 = 2$. Therefore,

$$\text{The Euler characteristic of the sphere} = 2. \tag{4.3.20}$$

In general, we investigate how the Euler characteristic of a surface changes when we attach a handle to the surface. We should attach a handle like Figure 4.3.10 but when we are calculating the Euler characteristic, the shape of the attached part may be considered as a triangle like Figure 4.3.11 as in the case where we considered a tetrahedron when calculating the Euler characteristic of the sphere. If we expand the handle H, it comes to be a sphere and so $\chi(H) = 2$. When H and S are separated, we have $\chi(S \cup H) = \chi(S) + \chi(H)$. When we attach H to S, we lose four triangles F_1, F_2, F_1', F_2' six edges of $\partial F_1', \partial F_2'$ are assimilated to edges of $\partial F_1, \partial F_2$ respectively, and six vertexes of F_1', F_2' are also assimilated to vertexes of F_1, F_2. Hence triangles, edges, and vertexes of the new surface S' is less than those of $S \cup H$ by 4,6,6 respectively. Therefore

$$\chi(S') = \chi(S) + \chi(H) - (6 - 6 + 4) = \chi(S) - 2. \tag{4.3.21}$$

Thus when we attach one handle, the Euler characteristic deceases by 2. If we attach a handle to the sphere and expand slightly the handle part, we obtain a torus like Figure 4.3.12. Moreover, if we attach one more handle to it, we obtain a surface like Figure 4.3.13. We call a surface obtained by attaching g handles a surface of **genus** g. The sphere is a surface of genus 0, the torus is a surface of genus 1. What we proved is that in general the Euler characteristic χ of a surface of genus g is given by

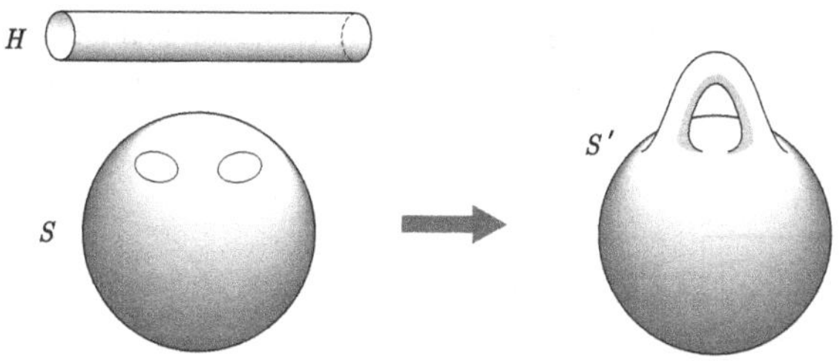

Fig. 4.3.10

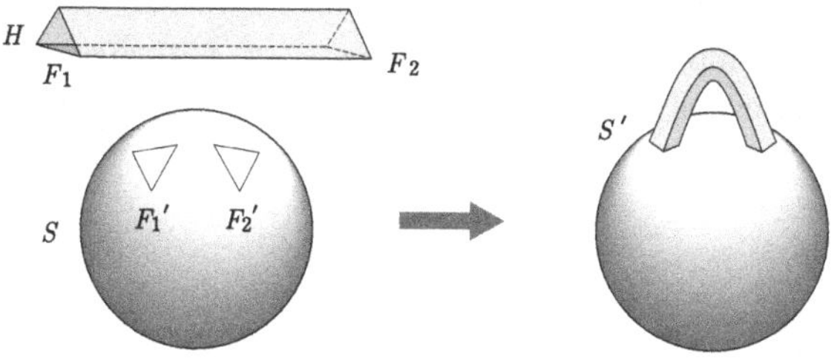

Fig. 4.3.11

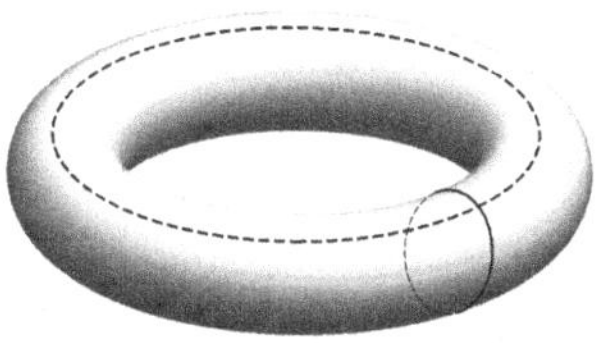

$g=1,\ \chi=0$

Fig. 4.3.12

$g=2,\ \chi=-2$

Fig. 4.3.13

$$\chi = 2 - 2g. \tag{4.3.22}$$

It is known that any oriented closed surface is obtained by attaching some handles to the sphere but we omit its proof because it is very long.

When we divide a torus in the space like Figure 4.3.12 into two parts by the circle on which z-coordinates attain the maximum and the circle on which z-coordinates attain the minimum, we have $K > 0$ on the outside part and $K < 0$ on the inside part. If we make a torus by pasting E_1 and E_1' together like Figure 4.3.14 and then pasting E_2 and E_2' together (not really but in your head), we naturally obtain the Riemannian metric of a torus from that of Euclidean plane. The curvature K is 0 and on the other hand the Gauss-Bonnet theorem concludes $\chi = 0$, which coincides to the fact we obtained by the above direct calculation. (If we cut a paper in a rectangle and paste E_1 and E_1' together, we can obtain a cylinder. However, if we then paste E_2 and E_2'

together, there must be places where $K > 0$ or $K < 0$, so it is not always $K = 0$. This is the reason why we should paste E_2 and E_2' together in your head.)

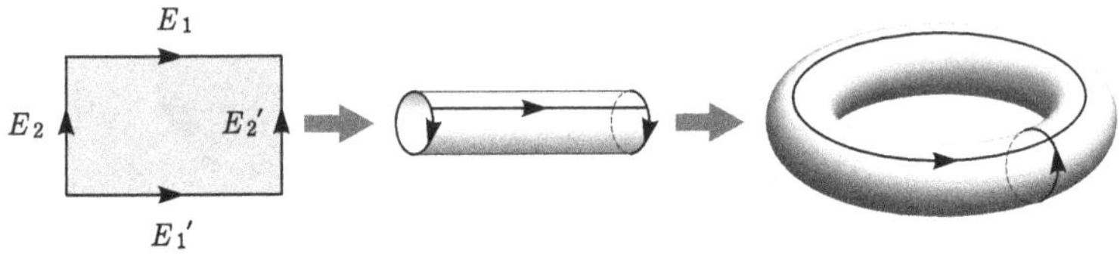

Fig. 4.3.14

It is known in complex analysis that there exists a Riemannian metric on an oriented closed surface of genus $g \geq 2$ so that $K = -1$. This also coincides with the Gauss-Bonnet theorem.

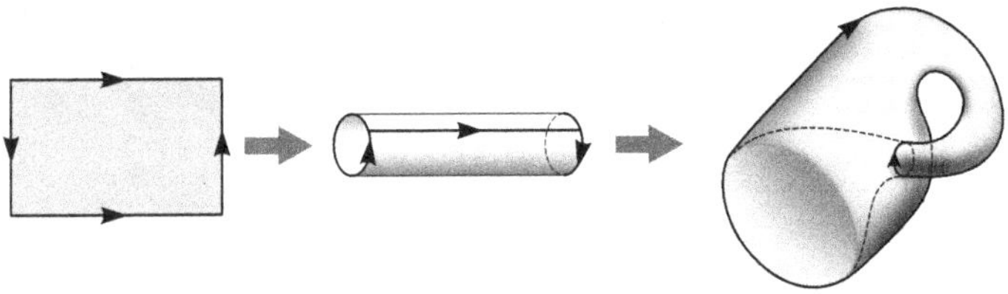

Fig. 4.3.15

Next, as an example of non-orientable closed surfaces we introduce the **Klein bottle**. It is obtained by pasting both sides of a cylinder together like Figure 4.3.15 along the opposite direction to the torus case in Figure 4.3.14. If we stretch or shrink some places, we get a strange bottle which has no interior. Usually an orientable surface has "the front side" and "the back side," but the Klein bottle has neither.

We assume that a surface S is given a triangulation and an orientation. We also assume that each edge of a triangle has an arrow like Figure 4.3.8. If we suppose that S is made of paper and we look at it from the back side, an arrow indicates the opposite direction, that is, if we go along the arrow we see the triangle on the right-hand side. This means that it is possible to distinguish clearly the front side and the back side of each triangle. Let us paint the front side with color red and the back side with color white. If S is oriented, since the arrows on the common edge of adjacent triangles indicate the opposite direction, a red triangle is next to a red triangle. Thus S is painted in red on one side and in white on the other. However we cannot define the orientation on the Klein bottle since if we move gradually on the surface, finally we reach the opposite side. (The above argument using the front side and the back side is not exact but it seems to be easy to understand intuitively.)

Problem 4.3.1 Let S be an oriented closed surface divided into some rectangles so that just four rectangles gather at each vertex (see Figure 4.3.16). Then, prove that

$$\int_S K\theta^1 \wedge \theta^2 = 0$$

for any Riemannian metric on S. (It is easy to see that a torus has such division conversely.)

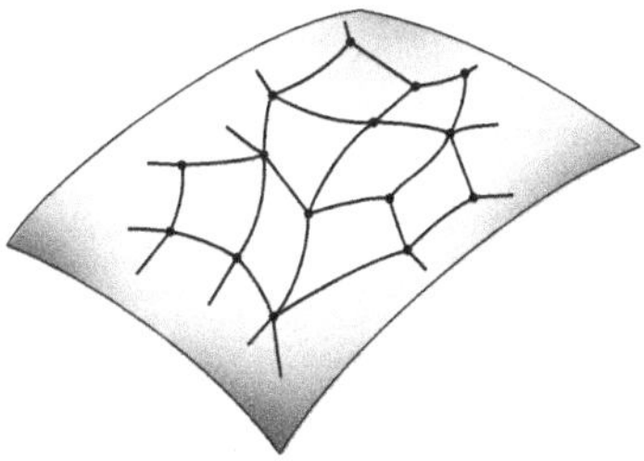

Fig. 4.3.16

Problem 4.3.2 Let φ be a differential 1-form defined on an oriented closed surface S. Prove

$$\int_S d\varphi = 0.$$

Problem 4.3.3 Let X be a vector field on an oriented closed surface S, which does not vanish anywhere. Prove that the Euler characteristic of S equals 0 by the Gauss-Bonnet formula and the formula in Problem 4.3.2.

Chapter 5
Minimal Surfaces

Although what we need was calculus and the elements of matrices as preliminary knowledge before this chapter, we need a little knowledge of complex function theory in this chapter. We may say that the interest of minimal surfaces lies in relation with complex function theory. In Section 2.3 of Chapter 2 we gave some problems about classical minimal surfaces. The aim of this chapter is to study much more about these surfaces. We do not mention at all questions like Plateau's problem which needs knowledge of difficult analysis. "A Survey of Minimal Surfaces" (1969) by R. Osserman is recommended to the readers who want to study much more about minimal surfaces.

5.1 Mean Curvatures and Minimal Surfaces

As we defined in Section 2.2 of Chapter 2, a minimal surface is a surface in the space whose mean curvature H is 0 everywhere. We explain the geometrical meaning of the condition $H = 0$.

We fix a domain R in the (u, v)-plane surrounded by a curve. We denote by ∂R the curve which is the boundary of R. Consider a surface

$$\boldsymbol{p}(u, v) = (x(u, v), y(u, v), z(u, v)) \tag{5.1.1}$$

where the parameter (u, v) varies in R. When (u, v) moves on ∂R, $\boldsymbol{p}(u, v)$ traces out a space curve. We deform the surface in fixing the space curve, i.e., we keep $\boldsymbol{p}(u, v)$ unchanged on ∂R. Although we explained that a surface never loses its area whatever we deform it satisfies the condition

$$H = 0$$

in Problem 2.2.4, we explain it a little in detail here with serving the review. Let $\boldsymbol{e}$ be a unit normal vector field and f be a function on R which takes the value 0 on the boundary ∂R. Then $f\boldsymbol{e}$ is a normal vector field on the surface and vanishes on the boundary. Thus if we set

The original version of the chapter was revised: Belated corrections have been incorporated. The correction to the chapter is available at https://doi.org/10.1007/978-981-15-1739-6_6

S. Kobayashi, *Differential Geometry of Curves and Surfaces*,
Springer Undergraduate Mathematics Series,
https://doi.org/10.1007/978-981-15-1739-6_5

$$\bar{\boldsymbol{p}} = \boldsymbol{p} + \varepsilon f \boldsymbol{e} \tag{5.1.2}$$

for a small real number ε, $\bar{\boldsymbol{p}}$ gives a surface which deforms $\boldsymbol{p}$ a little with keeping it fixed on ∂R. We denote by $A(\varepsilon)$ the area of the surface $\bar{\boldsymbol{p}}$. $A(0)$ is the area of the original surface $\boldsymbol{p}$. If we assume the area does not decrease when we deform $\boldsymbol{p}$, the inequality

$$A(0) \leq A(\varepsilon) \tag{5.1.3}$$

holds for every f and every ε. In particular, differentiating $A(\varepsilon)$ by ε and setting $\varepsilon = 0$, we get 0. That is,

$$A'(0) = 0. \tag{5.1.4}$$

On the other hand, as we showed in the solution of Problem 2.2.4,

$$A'(0) = -\iint_R fH\sqrt{EG - F^2}\,du\,dv. \tag{5.1.5}$$

If $H = 0$, the equation (5.1.4) holds clearly. Let φ be a function which is positive in R and 0 on ∂R. Setting $f = \varphi H$, we obtain

$$A'(0) = -\iint_R \varphi H^2\sqrt{EG - F^2}\,du\,dv \tag{5.1.6}$$

by (5.1.5). If $H \neq 0$, we get

$$A'(0) = -\iint_R \varphi H^2\sqrt{EG - F^2}\,du\,dv < 0. \tag{5.1.7}$$

Hence we find that the area is decreasing for sufficiently small ε if we deform the surface to the normal direction such as

$$\bar{\boldsymbol{p}} = \boldsymbol{p} + \varepsilon\varphi H\boldsymbol{e}.$$

In Problem 2.2.1 we confirmed that by setting

$$\begin{aligned} p &= \frac{\partial f}{\partial x}, & q &= \frac{\partial f}{\partial y}, & & \\ r &= \frac{\partial^2 f}{\partial x^2}, & s &= \frac{\partial^2 f}{\partial x \partial y}, & t &= \frac{\partial^2 f}{\partial y^2}, \end{aligned} \tag{5.1.8}$$

for a surface $z = f(x, y)$ we obtain

$$H = \frac{r(1 + q^2) - 2pqs + t(1 + p^2)}{2(1 + p^2 + q^2)^{3/2}}. \tag{5.1.9}$$

Hence the condition $H = 0$ in order to be a minimal surface is given by

$$r(1 + q^2) - 2pqs + t(1 + p^2) = 0. \tag{5.1.10}$$

It has been known as a differential equation of minimal surfaces since old times. However the equation which is useful in applications is the following **divergence form** rather than (5.1.9). That is, if we set

$$W = (1 + p^2 + q^2)^{\frac{1}{2}},$$

we get

$$H = \frac{1}{2}\left[\frac{\partial}{\partial x}\left(\frac{p}{W}\right) + \frac{\partial}{\partial y}\left(\frac{q}{W}\right)\right]. \tag{5.1.11}$$

We can prove it by reducing to (5.1.9) or by substituting

$$E = 1 + p^2, \qquad F = pq, \qquad G = 1 + q^2,$$
$$L = \frac{1}{W}\frac{\partial p}{\partial x}, \qquad M = \frac{1}{W}\frac{\partial q}{\partial x} = \frac{1}{W}\frac{\partial p}{\partial y}, \qquad N = \frac{1}{W}\frac{\partial q}{\partial y}$$

to (2.2.43) as Problem 2.2.1 we obtain

$$\begin{aligned}
H &= \frac{1}{2W^2}\left[\frac{1+q^2}{W}\frac{\partial p}{\partial x} + \frac{1+p^2}{W}\frac{\partial q}{\partial y} - \frac{pq}{W}\frac{\partial q}{\partial x} - \frac{pq}{W}\frac{\partial p}{\partial y}\right] \\
&= \frac{1}{2W^2}\left[\frac{W^2-p^2}{W}\frac{\partial p}{\partial x} + \frac{W^2-q^2}{W}\frac{\partial q}{\partial y} - \frac{pq}{W}\frac{\partial q}{\partial x} - \frac{pq}{W}\frac{\partial p}{\partial y}\right] \\
&= \frac{1}{2W^2}\left[\left\{W\frac{\partial p}{\partial x} - \frac{p}{W}\left(p\frac{\partial p}{\partial x} + q\frac{\partial q}{\partial x}\right)\right\} + \left\{W\frac{\partial q}{\partial y} - \frac{q}{W}\left(p\frac{\partial p}{\partial y} + q\frac{\partial q}{\partial y}\right)\right\}\right] \\
&= \frac{1}{2}\left[\frac{\partial}{\partial x}\left(\frac{p}{W}\right) + \frac{\partial}{\partial y}\left(\frac{q}{W}\right)\right].
\end{aligned}$$

5.2 Examples of Minimal Surfaces

We gave four examples of minimal surfaces in Section 2.3 of Chapter 2. Let us review them.

Example 5.2.1 Catenoid. By Example 2.3.8 this is a minimal surface obtained as a surface of revolution. It is obtained by rotating a catenary

$$x = a\cosh\frac{z}{a}$$

in the (x, z)-plane around the z-axis. By solving the equation of the catenary, we get

$$z = a\cosh^{-1}\frac{x}{a}.$$

From Example 2.1.8 we obtain the equation of the surface of revolution as

$$x = u\cos v, \qquad y = u\sin v, \qquad z = a\cosh^{-1}\frac{u}{a}.$$

Since $u = \sqrt{x^2 + y^2}$, it is expressed as

$$z = a\cosh^{-1}\frac{\sqrt{x^2 + y^2}}{a}.$$

Because the value of cosh is not less than 1, the equation above is defined on $x^2 + y^2 \geq a^2$. By rotating the curve in Figure 2.3.3 of Chapter 2 we obtain Figure 5.2.1. ♦

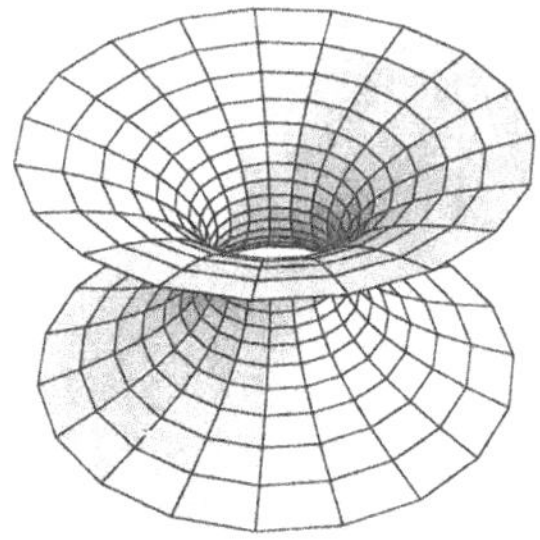

Fig. 5.2.1

Example 5.2.2 Right helicoid. This is a surface given by the parameter representation

$$x = u \cos v, \qquad y = u \sin v, \qquad z = av + b$$

as we showed Problem 2.3.2. If we write it as

$$\boldsymbol{p}(u, v) = (0, 0, av + b) + u(\cos v, \sin v, 0),$$

it is found that the surface is a ruled surface consisting of lines which pass $(0, 0, av + b)$ on the z-axis and whose direction $(\cos v, \sin v, 0)$ is orthogonal to the z-axis (see Problem 2.2.6). We can express the surface as

$$z = a \tan^{-1} \frac{y}{x} + b$$

by deleting u, v but it is not defined at $(0, 0)$ in this expression. A right helicoid is like Figure 5.2.2. ♦

Example 5.2.3 Scherk's minimal surface. In Problem 2.3.4 we showed Scherk's surface defined by

$$e^z \cos x = \cos y$$

is a minimal surface. Solving it in z, we write

$$z = \log \frac{\cos y}{\cos x}. \tag{5.2.1}$$

It is defined on the domain where $\cos x$ and $\cos y$ have the same signs. This domain consists of squares

$$S_{m,n} : |x - m\pi| < \frac{\pi}{2}, \quad |y - n\pi| < \frac{\pi}{2}, \tag{5.2.2}$$

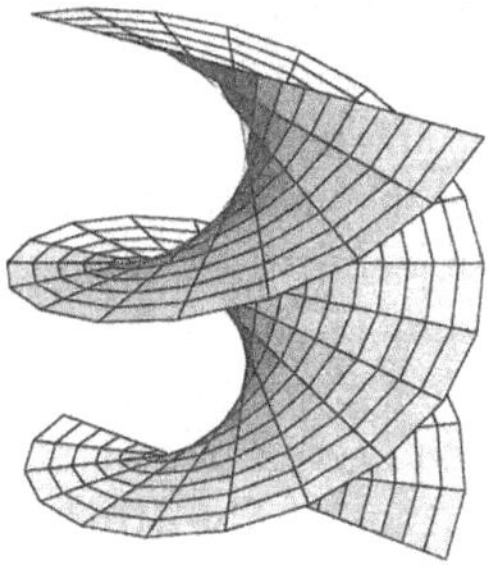

Fig. 5.2.2

where m, n are integers and $m + n$ is even, placed like a chessboard in Figure 5.2.3. The functions $\cos x, \cos y$ are periodic functions with period 2π and even functions as well. Thus by parallel displacing the graph on the square $S_{0,0}$, we obtain the surface.

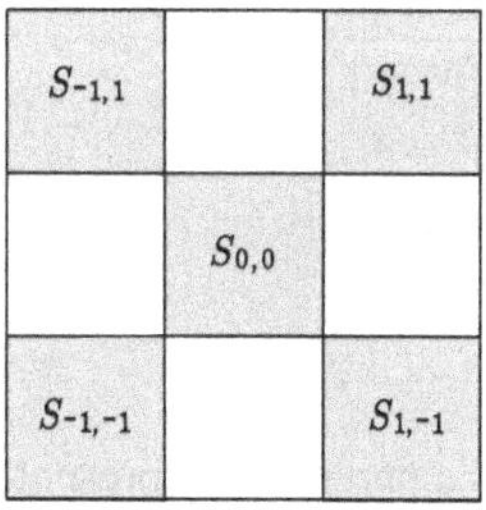

Fig. 5.2.3

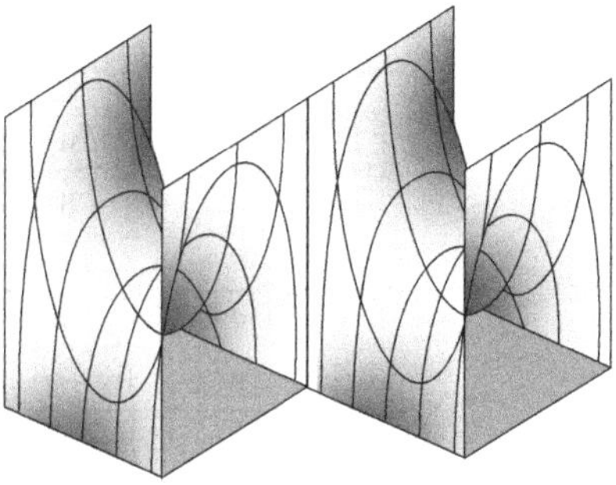

Fig. 5.2.4

Figure 5.2.4 shows the graph on one square. Since $\cos y = 0$ on the horizontal line of a square $S_{m,n}$, it implies $z = -\infty$. Since $\cos x = 0$ on the vertical line of a square $S_{m,n}$, it implies $z = +\infty$. When (x, y) approaches a vertex of the square, z can approach any value between $-\infty$ to $+\infty$ depending on the approaching direction.

For example, if we define θ by

$$\tan\theta = \frac{y - \frac{\pi}{2}}{x - \frac{\pi}{2}}$$

so that θ indicates the direction where (x, y) varies from $(0, 0)$ to $\left(\frac{\pi}{2}, \frac{\pi}{2}\right)$, As θ changes from 0 to $\frac{\pi}{2}$, the limit of z changes from $-\infty$ to $+\infty$. Hence the surface piece on

$S_{m,n}$ and that on $S_{m+1,n+1}$ commonly contains the vertical line at the common vertex. Although the surface has singularities on the vertical lines, which is because we use x, y as parameters, the surface is smooth everywhere including the vertical lines. In order to explain this fact we need the knowledge of analytic continuation so readers who are not familiar with the subject may skip the following part.

When we write a complex number α as $\alpha = |\alpha|e^{i\theta}$, we call θ as an **argument** of α and denote it by $\theta = \arg(\alpha)$. Since we have $e^{i\theta} = e^{i(\theta+2m\pi)}$ for an integer m, an argument is not defined uniquely. An argument is never defined for $\alpha = 0$. Arguments are defined uniquely not on $\mathbb{C} - \{0\}$ but on the universal covering of $\mathbb{C} - \{0\}$.

Now for a complex number $w = u + iv$ we define

$$x = \arg\frac{w+i}{w-i}, \qquad y = \arg\frac{w+1}{w-1}, \qquad z = \log\left|\frac{w^2+1}{w^2-1}\right|. \tag{5.2.3}$$

We cannot define (5.2.3) at four points $\pm 1, \pm i$. On the universal covering of $\mathbb{C} - \{\pm 1, \pm i\}$ x, y, z are uniquely defined. Let us show that these x, y, z satisfy (5.2.1). Since $\alpha = |\alpha|e^{i\theta}$ is written as $\alpha = |\alpha|(\cos\theta + i\sin\theta)$, we get

$$\cos(\arg\alpha) = \frac{1}{|\alpha|}\operatorname{Re}(\alpha),$$

where Re means the real part. Using this, we obtain

$$\begin{aligned} \cos x &= \left|\frac{w-i}{w+i}\right| \operatorname{Re}\left(\frac{w+i}{w-i}\right) = \frac{|w|^2-1}{|w+i||w-i|}, \\ \cos y &= \left|\frac{w-1}{w+1}\right| \operatorname{Re}\left(\frac{w+1}{w-1}\right) = \frac{|w|^2-1}{|w+1||w-1|}, \\ \frac{\cos y}{\cos x} &= \frac{|w+i|}{|w+1|}\frac{|w-i|}{|w-1|} = \left|\frac{w^2+1}{w^2-1}\right|. \end{aligned} \tag{5.2.4}$$

Thus it turns out that (5.2.1) is satisfied, hence (5.2.3) is a parameter representation of Scherk's surface.

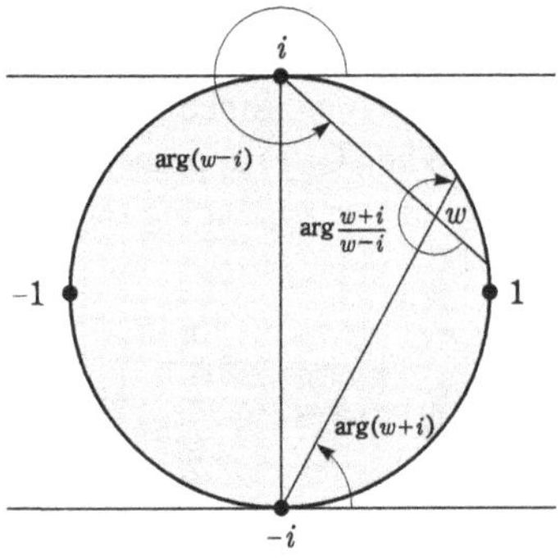

Fig. 5.2.5

The first two equations in (5.2.3) indicates that (x, y) are functions of w. It gives a one-to-one correspondence between the unit ball $|w| < 1$ and a quadrate domain $S_{m,n}$ where both m and n are odd. We explain it by using Figure 5.2.5. The vector pointing w from $-i$ is $w - (-i) = w + i$ and the vector pointing w from i is $w - i$. We define the arguments of these vectors under the condition

$$0 \leq \arg(w + i) < 2\pi, \qquad 0 \leq \arg(w - i) < 2\pi. \tag{5.2.5}$$

We define the argument $\arg\frac{w+i}{w-i}$ by

$$\arg\frac{w + i}{w - i} = \arg(w + i) - \arg(w - i). \tag{5.2.6}$$

Then it turns out by Figure 5.2.5 that

$$-\frac{3\pi}{2} < \arg\frac{w + i}{w - i} < -\frac{\pi}{2} \tag{5.2.7}$$

if $|w| < 1$. When we determine $\arg(w + 1)$ and $\arg(w - 1)$, we assign the condition

$$-\frac{\pi}{2} \leq \arg(w + 1) < \frac{3\pi}{2}, \qquad -\frac{\pi}{2} \leq \arg(w - 1) < \frac{3\pi}{2}. \tag{5.2.8}$$

(This is to avoid $\arg(w \pm 1)$ from changing discontinuously when w crosses the u-axis when w moves in the circle. However when w is in the unit circle, $\arg(w \pm 1)$ never takes the value $-\frac{\pi}{2}$.) By

$$\arg\frac{w + 1}{w - 1} = \arg(w + 1) - \arg(w - 1) \tag{5.2.9}$$

we get

$$-\frac{3\pi}{2} < \arg\frac{w + 1}{w - 1} < -\frac{\pi}{2}. \tag{5.2.10}$$

Hence when w moves in the unit circle, (x, y) moves in the quadrate domain $S_{-1,-1}$. This correspondence is one-to-one. (According to the theorem of Euclidean geometry, for a given argument x, w satisfying $x = \arg\frac{w+i}{w-i}$ lies on an arc of a circle through i and $-i$. Similarly, for a given y, w satisfying $y = \arg\frac{w+1}{w-1}$ lies on an arc of a circle through 1 and -1. Thus w determined by a given (x, y) is obtained as the intersection of these arcs of the circle.)

Since arg is not determined uniquely and has ambiguity of integer multiples of 2π, $S_{-1,-1}$ can be possibly replaced by $S_{m,n}$ (m, n: odd) depending on the choice of the arg.

We extend the map $w \mapsto (x, y) \in S_{-1,-1}$ defined above to the circle $|w| = 1$. First, x is not defined for $w = \pm i$. When w moves on a half circle $w = e^{it}$ $\left(-\frac{\pi}{2} < t < \frac{\pi}{2}\right)$, x is always $-\frac{3\pi}{2}$. When w skips i and moves on a half circle $w = e^{it}$ $\left(\frac{\pi}{2} < t < \frac{3\pi}{2}\right)$, x is always $-\frac{\pi}{2}$. Similarly, y is not defined for $w = \pm 1$. When w moves on a half circle $w = e^{it}$ $(0 < t < \pi)$, y is always $-\frac{\pi}{2}$. When w moves on a half circle $w = e^{it}$ $(\pi < t < 2\pi)$, y is always $-\frac{3\pi}{2}$. Hence on the four arcs

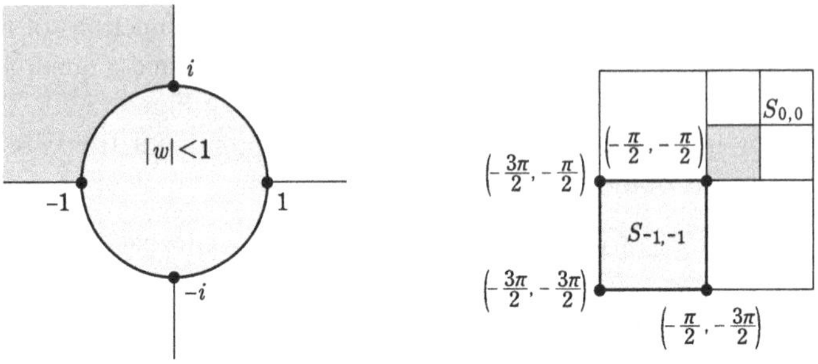

Fig. 5.2.6

$$w = e^{i\theta} \qquad \left(0 < t < \frac{\pi}{2}, \frac{\pi}{2} < t < \pi, \pi < t < \frac{3\pi}{2}, \frac{3\pi}{2} < t < 2\pi\right),$$

(x, y) is changed to the four vertexes of $S_{-1,-1}$

$$\left(-\frac{3\pi}{2}, -\frac{\pi}{2}\right), \left(-\frac{\pi}{2}, -\frac{\pi}{2}\right), \left(-\frac{\pi}{2}, -\frac{3\pi}{2}\right), \left(-\frac{3\pi}{2}, -\frac{3\pi}{2}\right)$$

respectively.

We extend the map $w \mapsto (x, y)$ further to the outside of the unit circle $|w| > 1$. Then how we choose the arg is a problem. We choose it so that $|w| > 1$ is mapped to $S_{0,0}$. Thus we extend the map beyond the circular arc

$$w = e^{it} \qquad \left(\frac{\pi}{2} < t < \pi\right)$$

which corresponds to the common vertex $\left(\frac{\pi}{2}, -\frac{\pi}{2}\right)$ of $S_{0,0}$ and $S_{-1,-1}$. That is, for w which lies on the second quadrant and satisfies $|w| > 1$, we determine

$$x = \arg\frac{w + i}{w - i}$$

by the conditions (5.2.5) and (5.2.6) and obtain $-\frac{\pi}{2} < x < 0$. Similarly, we determine y by the conditions (5.2.8) and (5.2.9) and obtain $-\frac{\pi}{2} < y < 0$. For w which lies on other quadrants we determine the arg so that the above map is extended continuously. As a result, we get the following correspondence:

$$|w| > 1 \text{ and } \begin{cases} \text{the first quadrant,} \\ \text{the second quadrant,} \\ \text{the third quadrant,} \\ \text{the fourth quadrant.} \end{cases} \Longrightarrow \quad (x, y) \in S_{0,0} \text{ and } \begin{cases} \text{the fourth quadrant,} \\ \text{the third quadrant,} \\ \text{the second quadrant,} \\ \text{the first quadrant.} \end{cases}$$

The infinity $w = \infty$ then corresponds to the origin $(0, 0)$ in $S_{0,0}$.

Now we have defined a map from $\mathbb{C} \cup \{\infty\}$ excluding four points $\pm 1, \pm i$ to $S_{-1,-1} \cup S_{0,0}$ with four vertexes of $S_{-1,-1}$. Except for on the unit circle $|w| = 1$ the map is one-to-one.

We denote by $M_{m,n}$ a part of Scherk's surface which is the graph over $S_{m,n}$. Figure 5.2.4 is one example of $M_{m,n}$. In order to investigate how $M_{-1,-1}$ and $M_{0,0}$ are connected, we observe how

$$z = \log \left| \frac{w^2 + 1}{w^2 - 1} \right|$$

moves when w moves on the circular arc

$$w = e^{i\theta} \qquad \left(\frac{\pi}{2} < t < \pi\right)$$

which corresponds to the common vertex $\left(-\frac{\pi}{2}, -\frac{\pi}{2}\right)$ of $S_{-1,-1}$ and $S_{0,0}$. Then

$$w^2 = e^{2i\theta} \qquad \left(\frac{\pi}{2} < t < \pi\right)$$

moves on the lower half circular arc from -1 to 1 and z increases monotonously from $-\infty$ to ∞. Therefore (x, y, z) moves from the lower $(-\infty)$ to the upper (∞) on the vertical line through $(x, y) = \left(-\frac{\pi}{2}, -\frac{\pi}{2}\right)$. The vertical line connects surfaces $M_{-1,-1}$ and $M_{0,0}$. What we have stated above will be clear when the readers study the Weierstrass-Enneper representation in Section 5.4. ♦

5.3 Isothermal Coordinate Systems

As we defined before Problem 3.2.1 of Chapter 3, we call a coordinate system (u, v) where a given Riemannian metric ds^2 on a surface is expressed as

$$ds^2 = E(du\,du + dv\,dv), \tag{5.3.1}$$

where E is a function of (u, v), an **isothermal coordinate system**. Although it is known that there exists such a coordinate system on a neighborhood of each point, its proof is difficult.

On a catenoid and a right helicoid which we treated in the previous section the first fundamental form is:

$$\mathbf{I} = du\,du + (u^2 + a^2)\,dv\,dv.$$

Thus (u, v) is not an isothermal coordinate but by coordinate transformation

$$\xi = \sinh^{-1} \frac{u}{a}, \qquad \eta = v,$$

the first fundamental form becomes

$$\mathbf{I} = a^2 \cosh^2 \xi\,(d\xi\,d\xi + d\eta\,d\eta)$$

and (ξ, η) is an isothermal coordinate system. In Problem 3.2.2 we got explicitly an isothermal coordinate system on a general surface of revolution. In this section, we will prove the existence of an isothermal coordinate system for minimal surfaces, since it a simple proof compared to that of a general surface.

Theorem 5.3.1 *Let M be a minimal surface in 3-dimensional Euclidean space. If we take an appropriately small neighborhood of any point of M, there exists an isothermal coordinate system on the neighborhood.*

Proof We may assume that M is given as

$$(x, y, f(x, y)), \qquad |x| < a, \qquad |y| < a,$$

locally. Then by using (5.1.11) we have

$$\begin{aligned}
&\frac{\partial}{\partial x}\left(-\frac{pq}{W}\right) + \frac{\partial}{\partial y}\left(\frac{1+p^2}{W}\right) \\
&= -q\frac{\partial}{\partial x}\left(\frac{p}{W}\right) - \frac{p}{W}\frac{\partial q}{\partial x} + \frac{\partial}{\partial y}\left(\frac{W^2 - q^2}{W}\right) \\
&= -q\left[\frac{\partial}{\partial x}\left(\frac{p}{W}\right) + \frac{\partial}{\partial y}\left(\frac{q}{W}\right)\right] - \frac{p}{W}\frac{\partial q}{\partial x} - \frac{q}{W}\frac{\partial q}{\partial y} + \frac{\partial W}{\partial y} \\
&= -2qH - \frac{p}{W}\frac{\partial p}{\partial y} - \frac{q}{W}\frac{\partial q}{\partial y} + \frac{\partial W}{\partial y} \\
&= -2qH = 0.
\end{aligned}$$

Similarly we have

$$\frac{\partial}{\partial x}\left(\frac{1+q^2}{W}\right) + \frac{\partial}{\partial y}\left(-\frac{pq}{W}\right) = -2pH = 0.$$

Hence the exterior derivatives of differential 1-forms

$$\alpha = \frac{1+p^2}{W}\,dx + \frac{pq}{W}\,dy, \qquad \beta = \frac{pq}{W}\,dx + \frac{1+q^2}{W}\,dy \tag{5.3.2}$$

vanish, i.e.,

$$d\alpha = 0, \qquad d\beta = 0.$$

By Poincaré's lemma (Theorem 5.1 of Chapter 2) there exists functions $\varphi(x, y)$, $\psi(x, y)$ defined on a domain $|x| < a$, $|y| < a$ such that

$$\alpha = d\varphi, \qquad \beta = d\psi. \tag{5.3.3}$$

If we set

$$u = x + \varphi(x, y), \qquad v = y + \psi(x, y), \tag{5.3.4}$$

then (u, v) is an isothermal coordinate system. In order to verify it, we calculate the transformation matrix. Since we have

$$\begin{bmatrix} \dfrac{\partial u}{\partial x} & \dfrac{\partial u}{\partial y} \\[2ex] \dfrac{\partial v}{\partial x} & \dfrac{\partial v}{\partial y} \end{bmatrix} = \begin{bmatrix} 1 + \dfrac{1+p^2}{W} & \dfrac{pq}{W} \\[2ex] \dfrac{pq}{W} & 1 + \dfrac{1+q^2}{W} \end{bmatrix}, \tag{5.3.5}$$

its determinant J is (by using $W^2 = 1 + p^2 + q^2$)

$$J = \frac{(W+1)^2}{W} > 0. \tag{5.3.6}$$

Hence we can express x, y as functions of u, v by the inverse function theorem. We obtain the inverse matrix of (5.3.5) by a simple calculation

$$\begin{bmatrix} \dfrac{\partial x}{\partial u} & \dfrac{\partial x}{\partial v} \\ \dfrac{\partial y}{\partial u} & \dfrac{\partial y}{\partial v} \end{bmatrix} = \begin{bmatrix} \dfrac{W+1+q^2}{JW} & -\dfrac{pq}{JW} \\ -\dfrac{pq}{JW} & \dfrac{W+1+p^2}{JW} \end{bmatrix}. \tag{5.3.7}$$

Therefore,

$$\begin{aligned} \frac{\partial z}{\partial u} &= \frac{\partial z}{\partial x}\frac{\partial x}{\partial u} + \frac{\partial z}{\partial y}\frac{\partial y}{\partial u} = \frac{p(W+1+q^2)}{JW} - \frac{pq^2}{JW} = \frac{p(W+1)}{JW}, \\ \frac{\partial z}{\partial v} &= \frac{\partial z}{\partial x}\frac{\partial x}{\partial v} + \frac{\partial z}{\partial y}\frac{\partial y}{\partial v} = -\frac{p^2q}{JW} + \frac{q(W+1+p^2)}{JW} = \frac{q(W+1)}{JW}. \end{aligned} \tag{5.3.8}$$

By (5.3.6), (5.3.7), and (5.3.8), we obtain the first fundamental form

$$\mathbf{I} = \left(\frac{W}{W+1}\right)^2 (du^2 + dv^2). \tag{5.3.9}$$

□

By using the isothermal coordinate system, the condition $H = 0$ of a minimal surface becomes simpler. First, we consider a general surface

$$\boldsymbol{p}(u, v) = (x(u, v),\ y(u, v),\ z(u, v)), \tag{5.3.10}$$

which is not necessarily a minimal surface. We assume that (u, v) is an isothermal coordinate system of the surface. Then $E = G,\ F = 0$ and we get

$$\boldsymbol{p}_u \cdot \boldsymbol{p}_u = \boldsymbol{p}_v \cdot \boldsymbol{p}_v, \qquad \boldsymbol{p}_u \cdot \boldsymbol{p}_v = 0. \tag{5.3.11}$$

By differentiating the first equation by u and the second equation by v,

$$\boldsymbol{p}_{uu} \cdot \boldsymbol{p}_u = \boldsymbol{p}_{vu} \cdot \boldsymbol{p}_v = -\boldsymbol{p}_u \cdot \boldsymbol{p}_{vv}.$$

Thus

$$(\boldsymbol{p}_{uu} + \boldsymbol{p}_{vv}) \cdot \boldsymbol{p}_u = 0. \tag{5.3.12}$$

Similarly, by differentiating the first equation of (5.3.11) by v and the second equation of (5.3.11) by u,

$$(\boldsymbol{p}_{uu} + \boldsymbol{p}_{vv}) \cdot \boldsymbol{p}_v = 0. \tag{5.3.13}$$

We write (5.3.12) and (5.3.13) as

$$\Delta \boldsymbol{p} \cdot \boldsymbol{p}_u = \Delta \boldsymbol{p} \cdot \boldsymbol{p}_v = 0 \tag{5.3.14}$$

by using Laplacian $\Delta = \frac{\partial^2}{\partial u^2} + \frac{\partial^2}{\partial v^2}$. Since $\boldsymbol{p}_u$ and $\boldsymbol{p}_v$ span the tangent space of the surface and (5.3.14) means $\Delta \boldsymbol{p}$ is orthogonal to the tangent space, $\Delta \boldsymbol{p}$ is a scalar multiple of the normal vector $\boldsymbol{e} = \frac{\boldsymbol{p}_u \times \boldsymbol{p}_v}{|\boldsymbol{p}_u \times \boldsymbol{p}_v|}$. The scalar is

$$\Delta \boldsymbol{p} \cdot \boldsymbol{e} = \boldsymbol{p}_{uu} \cdot \boldsymbol{e} + \boldsymbol{p}_{vv} \cdot \boldsymbol{e} = L + N \tag{5.3.15}$$

by (2.2.23).

On the other hand, since $H = \frac{N+L}{2E}$ by (2.2.44), we have

$$\Delta \boldsymbol{p} = 2EH\boldsymbol{e}. \tag{5.3.16}$$

Hence we have the following.

Theorem 5.3.2 *A surface $\boldsymbol{p}(u,v) = (x(u,v), y(u,v), z(u,v))$ parametrized by an isothermal coordinate system (u,v) is a minimal surface if and only if every component $x(u,v), y(u,v)$ or $z(u,v)$ is a harmonic function, i.e.,*

$$\Delta x(u,v) = \Delta y(u,v) = \Delta z(u,v) = 0.$$

Corollary 5.3.1 *There exists no compact minimal surface without boundary.*

Proof Let M be a compact minimal surface without boundary. If we use an isothermal coordinate system (u,v) on a neighborhood of a point where x is maximal, then $x(u,v)$ is a harmonic function. According to the maximal principle $x(u,v)$ is constant. A harmonic function which is constant on a nonempty open set is identically constant. Similarly y and z are also constant on M, which is a contradiction. □

5.4 The Weierstrass-Enneper Representation

For a given surface $\boldsymbol{p}(u,v) = (x(u,v), y(u,v), z(u,v))$ we consider complex functions

$$\varphi_1(w) = \frac{\partial x}{\partial u} - i\frac{\partial x}{\partial v}, \quad \varphi_2(w) = \frac{\partial y}{\partial u} - i\frac{\partial y}{\partial v}, \quad \varphi_3(w) = \frac{\partial z}{\partial u} - i\frac{\partial z}{\partial v} \tag{5.4.1}$$

of a variable $w = u + iv$. With simple calculation,

$$\varphi_1(w)^2 + \varphi_2(w)^2 + \varphi_3(w)^2 = E - G - 2iF, \tag{5.4.2}$$

$$|\varphi_1(w)|^2 + |\varphi_2(w)|^2 + |\varphi_3(w)|^2 = E + G. \tag{5.4.3}$$

Theorem 5.4.1 *For a given surface $\boldsymbol{p}(u,v) = (x(u,v), y(u,v), z(u,v))$ we define φ_1, φ_2, φ_3 by (5.4.1).*

(i) *$\varphi_1, \varphi_2, \varphi_3$ are holomorphic functions if and only if $x(u,v), y(u,v), z(u,v)$ are harmonic functions,*

(ii) *(u,v) is an isothermal coordinate system if and only if ${\varphi_1}^2 + {\varphi_2}^2 + {\varphi_3}^2 = 0$.*

Proof (i) In general $\zeta(w) = \xi(u,v) + i\eta(u,v)$ is holomorphic if and only if ξ and η satisfy the Cauchy-Riemann equations

$$\frac{\partial \xi}{\partial u} - \frac{\partial \eta}{\partial v} = 0, \qquad \frac{\partial \xi}{\partial v} + \frac{\partial \eta}{\partial u} = 0,$$

which we learned in complex function theory. Applying these to $\varphi_1(w)$, we obtain that $\varphi_1(w)$ is holomorphic if and only if

$$\frac{\partial^2 x}{\partial u^2} + \frac{\partial^2 x}{\partial v^2} = 0, \qquad \frac{\partial^2 x}{\partial u \partial v} - \frac{\partial^2 x}{\partial v \partial u} = 0.$$

Since the second equation always holds, it means that $x(u,v)$ is a harmonic function. Similarly $y(u,v), z(u,v)$ are also harmonic functions.

(ii) The condition for isothermal coordinate system is $E = G, F = 0$, which is implied from (5.4.2) immediately. □

Theorem 5.4.2 *Let $\mathbf{p}(u,v)$ be a surface defined on a domain D of the (u,v)-plane. We assume that $\mathbf{p}(u,v)$ is a minimal surface of which (u,v) is an isothermal coordinate system. Then $\varphi_1, \varphi_2, \varphi_3$ defined by (5.4.1) are holomorphic and satisfy ${\varphi_1}^2 + {\varphi_2}^2 + {\varphi_3}^2 = 0$. The first fundamental form is given by*

$$\mathbf{I} = \frac{1}{2}(|\varphi_1|^2 + |\varphi_2|^2 + |\varphi_3|^2)\, dw\, d\bar{w}.$$

Conversely, if holomorphic functions $\varphi_1, \varphi_2, \varphi_3$ on a simply connected domain D satisfy ${\varphi_1}^2 + {\varphi_2}^2 + {\varphi_3}^2 = 0$ and $|\varphi_1|^2 + |\varphi_2|^2 + |\varphi_3|^2 > 0$, then there exists a minimal surface $\mathbf{p}(u,v)$ which satisfies (5.4.1) and (u,v) is its isothermal coordinate system.

Proof The former part is obtained by Theorems 5.3.2 and 5.4.1. Since $\mathbf{p}(u,v)$ defines a surface, at least one of the determinants in (2.1.11) is not zero. Thus clearly $|\varphi_1|^2 + |\varphi_2|^2 + |\varphi_3|^2 > 0$. Because $E = G$, the first fundamental form is written as above by (5.4.3). Next, in order to prove the latter part we define

$$x = \mathrm{Re}\left(\int \varphi_1(w)\, dw\right), \quad y = \mathrm{Re}\left(\int \varphi_2(w)\, dw\right), \quad z = \mathrm{Re}\left(\int \varphi_3(w)\, dw\right).$$

Since D is simply connected, the above integrals can be defined. (Here these integrals mean $\int_{w_0}^{w}$ for arbitrary $w_0 \in D$.)

When we apply a differential operator

$$\frac{\partial}{\partial w} = \frac{1}{2}\left(\frac{\partial}{\partial u} - i\frac{\partial}{\partial v}\right)$$

to

$$2x = \int \varphi_1(w)\, dw + \int \overline{\varphi_1(w)}\, d\bar{w},$$

we get

$$\frac{\partial x}{\partial u} - i\frac{\partial x}{\partial v} = \varphi_1(w),$$

since

$$\frac{\partial}{\partial w}\int \overline{\varphi_1(w)}\, d\bar{w} = 0.$$

Similarly, we obtain the other two equations in (5.4.1). We will show that the $\boldsymbol{p}(u,v) = (x(u,v), y(u,v), z(u,v))$ we just obtained defines a surface. Take a point $w_0 = (u_0, v_0)$ in D. Since $|\varphi_1|^2 + |\varphi_2|^2 + |\varphi_3|^2 > 0$, we may assume $\varphi_1(w_0) \neq 0$. If all determinants in (2.1.11) are zero, there exists real numbers a and b such that $\varphi_2(w_0) = a\varphi_1(w_0)$ and $\varphi_3(w_0) = b\varphi_1(w_0)$. Thus we have

$$\varphi_1(w_0)^2 + \varphi_2(w_0)^2 + \varphi_3(w_0)^2 = (1 + a^2 + b^2)\varphi_1(w_0)^2 \neq 0,$$

which is a contradiction. Therefore at least one of determinants in (2.1.11) is not zero at (u_0, v_0). That is, $\boldsymbol{p}(u, v)$ defines a surface at (u_0, v_0). The rest is easily seen by Theorems 5.3.2 and 5.4.1. □

According to Theorem 5.4.2, it turns out that it is important to determine holomorphic functions $\varphi_1, \varphi_2, \varphi_3$ which satisfy

$$\varphi_1(w)^2 + \varphi_2(w)^2 + \varphi_3(w)^2 = 0 \tag{5.4.4}$$

when we study a minimal surface. The case where all of $\varphi_1, \varphi_2, \varphi_3$ are identically zero is excluded because it gives no surface. Hence we assume that φ_3 is not identically zero. Since $\varphi_1 \equiv i\varphi_2$ does not hold under this assumption, the set of points of D at which $\varphi_1 = i\varphi_2$ holds consists of isolated points only. If we define

$$f = \varphi_1 - i\varphi_2, \qquad g = \frac{\varphi_3}{\varphi_1 - i\varphi_2}, \tag{5.4.5}$$

then f is a holomorphic function and g is a meromorphic function (i.e., quotient of holomorphic functions). Since (5.4.4) implies

$$\varphi_1 + i\varphi_2 = -\frac{{\varphi_3}^2}{\varphi_1 - i\varphi_2} = -fg^2, \tag{5.4.6}$$

we obtain

$$\varphi_1 = \frac{1}{2}f(1-g^2), \quad \varphi_2 = \frac{i}{2}f(1+g^2), \quad \varphi_3 = fg. \tag{5.4.7}$$

Since $\varphi_1 + i\varphi_2$ in (5.4.6) is holomorphic, a point at which g has a pole of order m is a zero of order at least $2m$ of f. On the other hand, since

$$|\varphi_1|^2 + |\varphi_2|^2 + |\varphi_3|^2 = \frac{1}{2}|f|^2(1+|g|^2)^2,$$

a pole of order m of g must be a zero of just order $2m$ of f in order that $|\varphi_1|^2 + |\varphi_2|^2 + |\varphi_3|^2 > 0$. Altogether, we obtain the following Weierstrass-Enneper theorem.

Theorem 5.4.3 *A simply connected minimal surface is expressed as the following:*

$$p(w) = \left(\mathrm{Re}\int_0^w \frac{1}{2}f(1-g^2)\,dw,\ \mathrm{Re}\int_0^w \frac{i}{2}f(1+g^2)\,dw,\ \mathrm{Re}\int_0^w fg\,dw\right) \quad (w \in D),$$

where D is the unit open disk $\{w \in \mathbb{C} \mid |w| < 1\}$ or the total plane $\mathbb{C}$, $f(w)$ is a holomorphic function on D and $g(w)$ is a meromorphic function on D. The zeros *of $f(w)$ and the poles of $g(w)$ coincide and the order of zeros of $f(w)$ is twice of the order of poles of $g(w)$. Here* Re *means to take the real part. Moreover, the first fundamental form is given by*

$$\mathbf{I} = \frac{1}{4}|f|^2(1+|g|^2)^2\,dw\,d\bar{w}.$$

We note that D is the unit open disk or $\mathbb{C}$ because D is not the Riemann sphere by Corollary 5.3.3.

Example 5.4.1 Enneper's surface. In Theorem 5.4.3 let us set $D = \mathbb{C}, w = u + iv, f(w) = 2, g(w) = w$. Then we have

$$x = \operatorname{Re}\int_0^w (1-w^2)\,dw = \operatorname{Re}\left(w - \frac{1}{3}w^3\right) = u + uv^2 - \frac{1}{3}u^3,$$
$$y = \operatorname{Re}\int_0^w i(1+w^2)\,dw = \operatorname{Re}\left(i\left(w + \frac{1}{3}w^3\right)\right) = -v - u^2v + \frac{1}{3}v^3,$$
$$z = \operatorname{Re}\int_0^w 2w\,dw = \operatorname{Re}(w^2) = u^2 - v^2.$$

By multiplying these equations by three, we get the surface in Problem 2.3.3. ♦

Example 5.4.2 Scherk's surface. Now we observe Scherk's minimal surface again (see Example 5.2.3 and Problem 2.3.4). We use the expression given in (5.2.3)

$$x = \arg\frac{w+i}{w-i}, \quad y = \arg\frac{w+1}{w-1}, \quad z = \log\left|\frac{w^2+1}{w^2-1}\right| \quad (w = u+iv). \tag{5.4.8}$$

To differentiate the equations in (5.4.8) we write

$$\frac{w+i}{w-i} = \left|\frac{w+i}{w-i}\right| e^{ix}$$

and take the square of the both side

$$\left(\frac{w+i}{w-i}\right)^2 = \frac{w+i}{w-i}\frac{\bar{w}-i}{\bar{w}+i}e^{2ix}.$$

Then we get

$$\frac{w+i}{w-i} = \frac{\bar{w}-i}{\bar{w}+i}e^{2ix}. \tag{5.4.9}$$

We differentiate it by $w, \bar{w}$. Here the meaning of $\frac{\partial}{\partial w}, \frac{\partial}{\partial \bar{w}}$ is

$$\frac{\partial}{\partial w} = \frac{1}{2}\left(\frac{\partial}{\partial u} - i\frac{\partial}{\partial v}\right), \qquad \frac{\partial}{\partial \bar{w}} = \frac{1}{2}\left(\frac{\partial}{\partial u} + i\frac{\partial}{\partial v}\right).$$

Consequently, we obtain

$$\begin{aligned} \frac{-2i}{(w-i)^2} &= \frac{\bar{w}-i}{\bar{w}+i} e^{2ix} \cdot 2i \frac{\partial x}{\partial w}, \\ 0 &= \frac{2i}{(\bar{w}^2+i)^2} e^{2ix} + \frac{\bar{w}-i}{\bar{w}+i} e^{2ix} \cdot 2i \frac{\partial x}{\partial \bar{w}}. \end{aligned} \tag{5.4.10}$$

Arranging it by using (5.4.9),

$$\frac{\partial x}{\partial w} = \frac{-1}{w^2+1}, \qquad \frac{\partial x}{\partial \bar{w}} = \frac{-1}{\bar{w}^2+1}. \tag{5.4.11}$$

Similarly

$$\frac{\partial y}{\partial w} = \frac{i}{w^2-1}, \qquad \frac{\partial y}{\partial \bar{w}} = \frac{-i}{\bar{w}^2-1}. \tag{5.4.12}$$

We rewrite z as

$$z = \frac{1}{2} \log \frac{(w^2+1)(\bar{w}^2+1)}{(w^2-1)(\bar{w}^2-1)}$$

and differentiate it then we get

$$\frac{\partial z}{\partial w} = \frac{-2w}{w^4-1}, \qquad \frac{\partial z}{\partial \bar{w}} = \frac{-2\bar{w}}{\bar{w}^4-1}. \tag{5.4.13}$$

Therefore

$$dx = -\frac{dw}{w^2+1} - \frac{d\bar{w}}{\bar{w}^2+1}, \quad dy = \frac{idw}{w^2-1} - \frac{id\bar{w}}{\bar{w}^2-1}, \quad dz = -\frac{2wdw}{w^4-1} - \frac{2\bar{w}d\bar{w}}{\bar{w}^4-1}.$$

Hence the first fundamental form is

$$\mathbf{I} = dx^2 + dy^2 + dz^2 = \frac{4(|w|^2+1)^2}{|w^4-1|^2} dw\, d\bar{w}. \tag{5.4.14}$$

This shows that (u, v) is an isothermal coordinate system.

$\varphi_1, \varphi_2, \varphi_3$ defined by (5.4.1) are

$$\varphi_1 = 2\frac{\partial x}{\partial w} = \frac{-2}{w^2+1}, \quad \varphi_2 = 2\frac{\partial y}{\partial w} = \frac{2i}{w^2-1}, \quad \varphi_3 = 2\frac{\partial z}{\partial w} = \frac{-4w}{w^4-1} \tag{5.4.15}$$

by (5.4.11), (5.4.12) and (5.4.13). We obtain

$${\varphi_1}^2 + {\varphi_2}^2 + {\varphi_3}^2 = 0$$

by (5.4.15), while we also find it from Theorem 5.4.1 (ii). It is clear that $\varphi_1, \varphi_2, \varphi_3$ are holomorphic except for four points $\pm 1, \pm i$. They are holomorphic at the infinity $w = \infty$. Thus $x(u, v), y(u, v)$, and $z(u, v)$ are harmonic functions except for these four points. (This can be seen by Theorem 5.4.1 (i) and directly from (5.4.8) as well.)

Next we obtain functions f, g in (5.4.5) as

$$f = \varphi_1 - i\varphi_2 = \frac{4}{w^4-1}, \qquad g = \frac{\varphi_3}{\varphi_1 - i\varphi_2} = -w. \tag{5.4.16}$$

f is holomorphic except for $\pm 1, \pm i$ and g is a meromorphic function which has a pole at ∞. The Weierstrass-Enneper representation is

$$\begin{aligned} p(w) &= \left(\mathrm{Re}\int_0^w \frac{-2\,dw}{w^2+1}, \mathrm{Re}\int_0^w \frac{2i\,dw}{w^2-1}, \mathrm{Re}\int_0^w \frac{-4w\,dw}{w^4-1}\right) \\ &= \left(\mathrm{Re}\left(-i\log\frac{w+i}{w-i}\right), \mathrm{Re}\left(-i\log\frac{w+1}{w-1}\right), \mathrm{Re}\left(\log\frac{w^2+1}{w^2-1}\right)\right). \end{aligned} \tag{5.4.17}$$

However this is not defined uniquely on $\mathbb{C}\cup\{\infty\}-\{\pm1,\pm i\}$ and we should consider that this is defined on its universal covering. Of course (5.4.17) and (5.4.8) coincide. ♦

Example 5.4.3 Catenoid and right helicoid. As we see in Section 2.3 of Chapter 2 they have the same first fundamental form so we explain these surfaces together. We consider holomorphic functions

$$\varphi_1 = 2e^{it}\cosh w, \qquad \varphi_2 = -2ie^{it}\sinh w, \qquad \varphi_3 = -2ie^{it} \tag{5.4.18}$$

of $w = u + iv$ for each real number $0 \leq t < 2\pi$. Clearly ${\varphi_1}^2 + {\varphi_2}^2 + {\varphi_3}^2 = 0$. We define

$$x = \frac{1}{2}\mathrm{Re}\int_0^w \varphi_1 dw, \qquad y = \frac{1}{2}\mathrm{Re}\int_0^w \varphi_2 dw, \qquad z = \frac{1}{2}\mathrm{Re}\int_0^w \varphi_3 dw$$

so that (5.4.1) satisfies. Then

$$\begin{aligned} x &= \mathrm{Re}(e^{it}\sinh w) = \frac{1}{2}\mathrm{Re}(e^u e^{i(v+t)} - e^{-u}e^{i(-v+t)}) \\ &= \frac{1}{2}\mathrm{Re}[e^u\{\cos(v+t) + i\sin(v+t)\} - e^{-u}\{\cos(-v+t) + i\sin(-v+t)\}] \\ &= \frac{1}{2}[e^u(\cos v\cos t - \sin v\sin t) - e^{-u}(\cos v\cos t + \sin v\sin t)] \\ &= \sinh u\cos v\cos t - \cosh u\sin v\sin t. \end{aligned}$$

Similar calculation gives y and z. Altogether, we get

$$\begin{aligned} x &= \sinh u\cos v\cos t - \cosh u\sin v\sin t, \\ y &= \sinh u\sin v\cos t + \cosh u\cos v\sin t, \\ z &= v\cos t + u\sin t. \end{aligned} \tag{5.4.19}$$

For each t, (5.4.19) is the Weierstrass-Enneper representation which defines a minimal surface.

In the case of $t = 0$, we have

$$x = \sinh u\cos v, \qquad y = \sinh u\sin v, \qquad z = v. \tag{5.4.20}$$

Setting $\xi = \sinh u, \eta = v$, we get

$$x = \xi\cos\eta, \qquad y = \xi\sin\eta, \qquad z = \eta,$$

which is the equation of the right helicoid in Example 5.2.2.

In the case of $t = \frac{\pi}{2}$, we have

$$x = -\cosh u\sin v, \qquad y = \cosh u\cos v, \qquad z = u. \tag{5.4.21}$$

Setting $\xi = \cosh u, \eta = -v$, we get

$$x = \xi \sin \eta, \qquad y = \xi \cos \eta, \qquad z = \cosh^{-1} \xi,$$

which is the equation of the catenoid in Example 5.2.1 (note that x and y are exchanged). In this way we obtain the right catenoid and the helicoid as special cases in the family of minimal surfaces parametrized by t.

(5.4.18) implies

$$\begin{aligned} f &= \varphi_1 - i\varphi_2 = 2e^{it}\, e^{-w}, \\ g &= \frac{\varphi_3}{\varphi_1 - i\varphi_2} = -i\, e^{w}. \end{aligned} \tag{5.4.22}$$

We should note that in the case of the catenoid, i.e., $t = \frac{\pi}{2}$ is slightly different. If we change v in (5.4.21) to $v + 2ki\pi$, x, y, z do not change. It leads that the Weierstrass-Enneper representation $\boldsymbol{p} : \mathbb{C} \to \mathbb{R}^3$ is a covering map where $\mathbb{C}$ is the universal covering of the catenoid. On the other hand, for other t $(0 \leq t < 2\pi)$ including right helicoid, the Weierstrass-Enneper representation gives a one-to-one correspondence between $\mathbb{C}$ and the surface. ♦

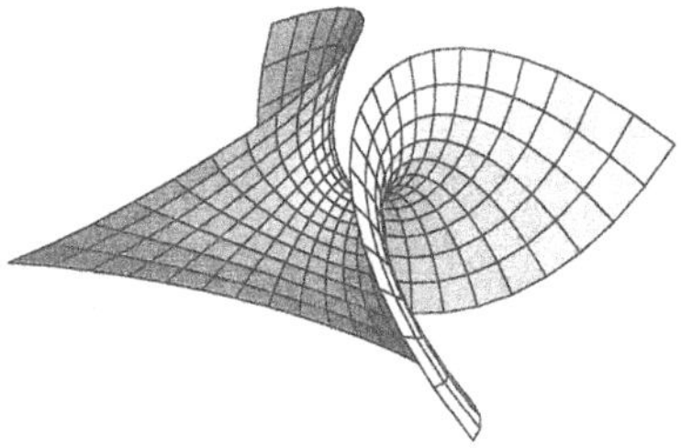

Fig. 5.4.1

5.5 Associated Minimal Surfaces

When a minimal surface is given by the Weierstrass-Enneper representation $\boldsymbol{p} : D \to \mathbb{R}^3$

$$\boldsymbol{p}(w) = \left(\mathrm{Re} \int_0^w \frac{1}{2} f(1 - g^2) dw,\ \mathrm{Re} \int_0^w \frac{i}{2} f(1 + g^2) dw,\ \mathrm{Re} \int_0^w f g dw \right)$$

as stated in Theorem 5.4.3, we set $f_\theta = e^{i\theta} f$ for arbitrary θ $(0 \leq \theta < \pi)$ and define $\boldsymbol{p}_\theta : D \to \mathbb{R}^3$ as

$$p_\theta(w) = \left(\mathrm{Re}\int_0^w \frac{1}{2} f_\theta(1-g^2)\,dw,\ \mathrm{Re}\int_0^w \frac{i}{2} f_\theta(1+g^2)\,dw,\ \mathrm{Re}\int_0^w f_\theta g\,dw\right). \tag{5.5.1}$$

Then a surface $p_\theta(D)$ in $\mathbb{R}^3$ changes together with θ but we find that the first fundamental form defined on D from p_θ is invariant by Theorem 5.4.3. These minimal surfaces p_θ obtained like this are called **associated minimal surfaces** of p. In particular, we call $p_{\frac{\pi}{2}}$ the **conjugate minimal surface** of p.

Since

$$p(w) - i p_{\frac{\pi}{2}}(w) = \left(\int_0^w \frac{1}{2} f(1-g^2)\,dw,\ \int_0^w \frac{i}{2} f(1+g^2)\,dw,\ \int_0^w f g\,dw\right), \tag{5.5.2}$$

if we set

$$p_\theta(w) = (x_\theta(u,v),\ y_\theta(u,v),\ z_\theta(u,v)),$$

$-x_{\frac{\pi}{2}}(u,v), -y_{\frac{\pi}{2}}(u,v), -z_{\frac{\pi}{2}}(u,v)$ are conjugate harmonic functions of $x_0(u,v)$, $y_0(u,v)$, $z_0(u,v)$, respectively.

Example 5.5.1 Right helicoid and catenoid. Let a be a positive real number and we define

$$f(w) = ae^w, \qquad g(w) = ie^{-w}$$

for $w \in \mathbb{C}$. If we set $w = u + iv$,

$$\begin{aligned}
\int_0^w \frac{1}{2} f(1-g^2)dw &= \int_0^w \frac{a}{2}(e^w + e^{-w})dw \\
&= \frac{a}{2}(e^w - e^{-w}) \\
&= \frac{a}{2}(e^u e^{iv} - e^{-u} e^{-iv}) \\
&= \frac{a}{2}\{e^u(\cos v + i\sin v) - e^{-u}(\cos v - i\sin v)\} \\
&= a\sinh u\cos v + ia\cosh u\sin v.
\end{aligned}$$

Similarly

$$\begin{aligned}
&\int_0^w \frac{i}{2} f(1+g^2)dw = -a\sinh u\sin v + ia\cosh u\cos v, \\
&\int_0^w fg dw = iaw = -av + iau.
\end{aligned}$$

Taking real parts of them, we obtain a minimal surface

$$p(u,v) = (a\sinh u\cos v, -a\sinh u\sin v, -av). \tag{5.5.3}$$

Moreover,

$$p_{\frac{\pi}{2}}(u,v) = (-a\cosh u\sin v, -a\cosh u\cos v, -au) \tag{5.5.4}$$

gives the conjugate minimal surface of (5.5.3).

By changing variables of a surface (5.5.3) as

$$\xi = a \sinh u, \qquad \eta = -v, \tag{5.5.5}$$

we get

$$\boldsymbol{p}(u(\xi,\eta), v(\xi,\eta)) = (\xi \cos\eta,\ \xi \sin\eta,\ a\eta). \tag{5.5.6}$$

It is a right helicoid in Examples 5.2.2 and Problem 2.3.2. By the same change of variables as (5.5.5) the surface (5.5.4) is expressed as

$$\boldsymbol{p}_{\frac{\pi}{2}}(u(\xi,\eta), v(\xi,\eta)) = \left(\sqrt{\xi^2 + a^2}\sin\eta,\ -\sqrt{\xi^2 + a^2}\cos\eta,\ -a\sinh^{-1}\frac{\xi}{a}\right). \tag{5.5.7}$$

It is nothing but a catenoid in Examples 5.2.1 and 2.3.8. Therefore the right helicoid and the catenoid are conjugate. ♦

5.6 Curvatures of Minimal Surfaces

When a minimal surface $\boldsymbol{p}(u,v) = (x(u,v), y(u,v), z(u,v))$ is represented by an isothermal coordinate system, we define in (5.4.2) and (5.4.5) holomorphic functions $\varphi_1(w)$, $\varphi_2(w)$, $\varphi_3(w)$, $f(w)$ and a meromorphic function $g(w)$ as

$$\varphi_1(w) = \frac{\partial x}{\partial u} - i\frac{\partial x}{\partial v}, \qquad \varphi_2(w) = \frac{\partial y}{\partial u} - i\frac{\partial y}{\partial v}, \qquad \varphi_3(w) = \frac{\partial z}{\partial u} - i\frac{\partial z}{\partial v},$$
$$f = \varphi_1 - i\varphi_2, \qquad g = \frac{\varphi_3}{\varphi_1 - i\varphi_2}.$$

Hence

$$\begin{aligned}
\boldsymbol{p}_u &= \frac{1}{2}(\varphi_1 + \bar{\varphi}_1,\ \varphi_2 + \bar{\varphi}_2,\ \varphi_3 + \bar{\varphi}_3),\\
\boldsymbol{p}_v &= \frac{i}{2}(\varphi_1 - \bar{\varphi}_1,\ \varphi_2 - \bar{\varphi}_2,\ \varphi_3 - \bar{\varphi}_3),\\
\boldsymbol{p}_u \times \boldsymbol{p}_v &= \mathrm{Im}(\varphi_2\bar{\varphi}_3,\ \varphi_3\bar{\varphi}_1,\ \varphi_1\bar{\varphi}_2) \qquad \text{(Im means the imaginary part)}\\
&= \frac{|f|^2(1+|g|^2)}{4}(2\mathrm{Re}(g), 2\mathrm{Im}(g), |g|^2 - 1).
\end{aligned} \tag{5.6.1}$$

Here we use (5.4.7) in the last equality. Thus a unit normal vector $\boldsymbol{e}$ is given by

$$\boldsymbol{e} = \frac{\boldsymbol{p}_u \times \boldsymbol{p}_v}{|\boldsymbol{p}_u \times \boldsymbol{p}_v|} = \left(\frac{2\,\mathrm{Re}(g)}{|g|^2 + 1}, \frac{2\,\mathrm{Im}(g)}{|g|^2 + 1}, \frac{|g|^2 - 1}{|g|^2 + 1}\right). \tag{5.6.2}$$

The first fundamental form is

$$\mathbf{I} = \frac{1}{2}(|\varphi_1|^2 + |\varphi_2|^2 + |\varphi_3|^2)dw d\bar{w}$$

by Theorem 5.4.2. Using (5.4.7), we get

$$\mathbf{I} = \left(\frac{|f|(|g|^2 + 1)}{2}\right)^2 dw\, d\bar{w}. \tag{5.6.3}$$

By (5.6.1) and

$$\frac{\partial}{\partial w} = \frac{1}{2}\left(\frac{\partial}{\partial u} - i\frac{\partial}{\partial v}\right), \qquad \frac{\partial}{\partial \bar{w}} = \frac{1}{2}\left(\frac{\partial}{\partial u} + i\frac{\partial}{\partial v}\right),$$

we obtain

$$\begin{aligned} \boldsymbol{p}_{uu} &= \frac{1}{2}(\varphi_1{}' + \bar{\varphi}_1{}', \varphi_2{}' + \bar{\varphi}_2{}', \varphi_3{}' + \bar{\varphi}_3{}'), \\ \boldsymbol{p}_{uv} = \boldsymbol{p}_{vu} &= \frac{i}{2}(\varphi_1{}' - \bar{\varphi}_1{}', \varphi_2{}' - \bar{\varphi}_2{}', \varphi_3{}' - \bar{\varphi}_3{}'), \\ \boldsymbol{p}_{vv} &= -\frac{1}{2}(\varphi_1{}' + \bar{\varphi}_1{}', \varphi_2{}' + \bar{\varphi}_2{}', \varphi_3{}' + \bar{\varphi}_3{}'). \end{aligned}$$

Calculating L, M, N by using (5.4.7), (5.6.2) and (2.2.23), we get

$$L = -N = -\mathrm{Re}(fg'), \qquad M = \mathrm{Im}(fg'). \tag{5.6.4}$$

Hence the second fundamental form is given by

$$\mathbf{II} = -\mathrm{Re}\left(f\frac{dg}{dw}\,dw^2\right). \tag{5.6.5}$$

We calculate the Gaussian curvature K by (2.2.43) and obtain principal curvatures κ_1, κ_2 by $H = 0$:

$$\begin{aligned} \kappa_1 &= \frac{4\,|g'|}{|f|(|g|^2+1)^2}, \qquad \kappa_2 = \frac{-4\,|g'|}{|f|(|g|^2+1)^2}, \\ K &= -\left(\frac{4\,|g'|}{|f|(|g|^2+1)^2}\right)^2. \end{aligned} \tag{5.6.6}$$

Hence if we define

$$d\tilde{s}^2 = \sqrt{-K}\mathbf{I},$$

we get

$$d\tilde{s}^2 = |f|\,|g'|\,dwd\bar{w}.$$

Therefore it turns out that the Gaussian curvature of the Riemannian metric $d\tilde{s}^2$ is zero by (3.2.43). The converse is also known:

Theorem 5.6.1 *Let (M, ds^2) be a 2-dimensional simply connected Riemannian manifold. We assume that its Gaussian curvature K is negative. Then there exists a minimal surface $\boldsymbol{p} : M \to \mathbb{R}^3$ such that $\mathbf{I} = d\boldsymbol{p} \cdot d\boldsymbol{p} = ds^2$ if and only if the Gaussian curvature of $d\tilde{s}^2 = \sqrt{-K}\mathbf{I}$ is zero.*

This theorem was proved by Ricci. We gave the proof of the if part. Since the proof of the only-if part is difficult, we omit it.

5.7 Gauss' Spherical Maps

When a minimal surface $\boldsymbol{p} : D \longrightarrow \mathbb{R}^3$ is given by the Weierstrass-Enneper representation as in Theorem 5.4.3, we will observe the Gauss map (see Section 2.2 of Chapter 2) $\boldsymbol{e} : D \to \mathbb{R}^3$. Here $\boldsymbol{e}$ is the unit normal vector given by (5.6.2).

First we need to explain the **stereographic projection**. Let S^2 be the unit sphere in $\mathbb{R}^3$ whose center is the origin. We give the following correspondence between a point $(x, y, z) \neq (0, 0, 1)$ of S^2 and a point $\zeta = \xi + i\eta$ in $\mathbb{C}$. Let $(\xi, \eta, 0)$ be the intersection point of the line

$$t(x, y, z) + (1 - t)(0, 0, 1) = (tx, ty, t(z - 1) + 1),$$

which joins the north pole $N = (0, 0, 1)$ and (x, y, z), and the plane $\mathbb{C}$ like Figure 5.7.1. Since $t = \frac{1}{1-z}$ at the point, we get

$$\xi = \frac{x}{1 - z}, \qquad \eta = \frac{y}{1 - z}. \tag{5.7.1}$$

Noting $x^2 + y^2 + z^2 = 1$, we solve (5.7.1) then we get

$$x = \frac{2\xi}{|\zeta|^2 + 1}, \qquad y = \frac{2\eta}{|\zeta|^2 + 1}, \qquad z = \frac{|\zeta|^2 - 1}{|\zeta|^2 + 1}. \tag{5.7.2}$$

In this way we obtain a one-to-one correspondence between S^2 and $\mathbb{C} \cup \{\infty\}$. Here we define that the north pole N of S^2 corresponds to the infinity ∞.

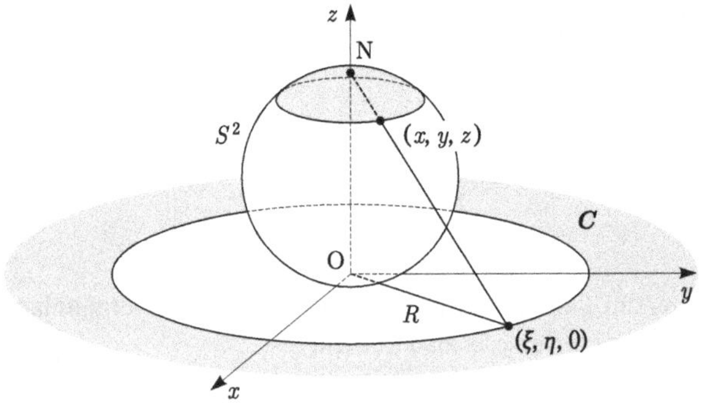

Fig. 5.7.1

On the other hand, $\boldsymbol{e}$ is given by (5.6.2) as follows:

$$\boldsymbol{e} = \left(\frac{2\,\mathrm{Re}(g)}{|g|^2 + 1}, \frac{2\,\mathrm{Im}(g)}{|g|^2 + 1}, \frac{|g|^2 - 1}{|g|^2 + 1} \right). \tag{5.7.3}$$

Comparing (5.7.2) and (5.7.3), we find the following. The composition of the Gauss map $\boldsymbol{e} : D \longrightarrow S^2$ and the stereographic projection $S^2 \longrightarrow \mathbb{C} \cup \{\infty\}$ is the map g.

Let a surface $\boldsymbol{p} : D \longrightarrow \mathbb{R}^3$ defined on a domain D in the (u, v)-plane be a minimal surface such that (u, v) is its isothermal coordinate system. If D is not simply connected, we take the universal covering $\tilde{D}$ (or simply connected subdomain of D) and apply the argument above by using the Weierstrass-Enneper representation. Altogether, we obtain the following theorem.

Theorem 5.7.1 *Let $\boldsymbol{p} : D \longrightarrow \mathbb{R}^3$ be a surface defined on the (u, v)-plane such that (u, v) is its isothermal coordinate system. If the surface is a minimal surface, the Gauss map is given by a meromorphic function g on $\tilde{D}$ under the identification of S^2 and $\mathbb{C} \cup \{\infty\}$.*

The value of $\boldsymbol{e}$ is the north pole $(0, 0, 1)$ at a pole of g in (5.7.3).

The converse of Theorem 5.7.1 is also known. That is, if the Gauss map is a holomorphic map form D to $\mathbb{C} \cup \{\infty\}$, then a surface $\boldsymbol{p}$ is a minimal surface. We omit the proof.

Example 5.7.1 Enneper's surface. The image of the Gauss map of Enneper's surface which was defined in Problem 2.3.3 covers S^2 just once except for the north pole as we explained in Problem 2.3.6 (i). We can also find it by $g(w) = w$ in the Weierstrass-Enneper representation shown in Example 5.4.1. ♦

Example 5.7.2 Catenoid and right helicoid. As we explained in Problem 2.3.6 (g) and (h), the image of the Gauss map covers S^2 except for the north pole and the south pole in both cases. We can also find it by $g(w) = -ie^w$ in the Weierstrass-Enneper representation shown in Example 5.4.3. As we also explained in Problem 2.3.6, it covers just once in case of a catenoid and infinite times in case of the right helicoid. Why is there a difference? As $g(w + 2k\pi i) = g(w)$, it seems that the image of the Gauss map covers S^2 except for both poles infinite times in both cases. However w and $w + 2k\pi i$ gives the same point on the catenoid as explained in Example 5.4.3 hence we should limit the range of w to $0 \leq v < 2\pi$ in case of a catenoid. ♦

Example 5.7.3 Scherk's surface. As we explained in detail in Example 5.2.3, in Scherk's surface defined in Problem 2.3.4 $M_{m,n}$ and $M_{m+1,n+1}$ are connected by the vertical line at the common vertex of $S_{m,n}$ and $S_{m+1,n+1}$, where $M_{m,n}$ is a part of the surface defined on a square domain $S_{m,n}$. As we explained in Problem 2.3.6 (j), $M_{0,0}$ covers just once the upper semisphere (not including the equator) by the Gauss map. We can also find it by $g(w) = -w$ which is induced from the Weierstrass-Enneper representation shown in Example 5.4.2. ($|w| > 1$ corresponded to $M_{0,0}$ there.) Since $M_{-1,-1}$ is defined on $|w| < 1$, $M_{-1,-1}$ covers just once the lower semisphere by the Gauss map. Although $|w| = 1$ corresponds to the equator, $g(w)$ does not take $\pm 1, \pm i$ since Scherk's surface does not defined at $w = \pm 1, \pm i$. The corresponding four points on the equator are $(\pm 1, \pm 1, 0)$. The vertical lines at four vertexes of $S_{0,0}$ correspond to four arcs of circle which is obtained by excluded $\pm 1, \pm i$ from the circle $|w| = 1$ and four arcs of circle excluded four exceptional points. Therefore the image of the Gauss map covers S^2 infinite times except for only four points $(\pm 1, \pm 1, 0)$ on the equator. ♦

Generally, we consider the Weierstrass-Enneper representation $\boldsymbol{p} : D \longrightarrow \mathbb{R}^3$ of a given minimal surface. (In order to let D be simply connected, we take the universal covering of D if we need.) Riemann's theorem states that a simply connected domain D is analytically (holomorphically) isomorphic to $\mathbb{C}$ or the unit disc of $\mathbb{C}$. When $D = \mathbb{C}$, Picard's little theorem leads that a holomorphic map $g : \mathbb{C} \to \mathbb{C} \cup \{\infty\}$ is constant if g has at least three exceptional values. Moreover, if the Gauss map is constant, the surface is actually a plane. As for the number of exceptional values of the Gauss map of a minimal surface, refer to the following theorem by Hirotaka Fujimoto (H. Fujimoto, On the number of exceptional values of the Gauss maps of minimal surfaces, J. Math. Soc. Japan 40 (1988), 235–247).

Theorem 5.7.2 *The number of exceptional values of the Gauss map of a non-flat complete minimal surface is at most four.*

Since Scherk's surface has just four exceptional values, this is the best possible result.

Correction to: Differential Geometry of Curves and Surfaces

Correction to:
S. Kobayashi, *Differential Geometry of Curves and Surfaces*, Springer Undergraduate Mathematics Series, https://doi.org/10.1007/978-981-15-1739-6

The original version of the book was inadvertently published with errors in equations and text on the following pages: 009, 013, 017, 020, 026, 062, 071, 104, 121, 124, 131, 133, 139, and 148. These have now been updated and approved by the author.

The updated version of these chapters can be found at
https://doi.org/10.1007/978-981-15-1739-6_1
https://doi.org/10.1007/978-981-15-1739-6_2
https://doi.org/10.1007/978-981-15-1739-6_3
https://doi.org/10.1007/978-981-15-1739-6_4
https://doi.org/10.1007/978-981-15-1739-6_5

S. Kobayashi, *Differential Geometry of Curves and Surfaces*,
Springer Undergraduate Mathematics Series,
https://doi.org/10.1007/978-981-15-1739-6_6

Appendix

Theorem of differential equations used in the text

We will state a theorem on the existence and uniqueness of the solution of an ordinary differential equation which was used in surface theory, parallel transformations, geodesics and so on. As for the proof, please refer to suitable textbooks on differential equations.

Theorem *Let $f^i(t; y^1, \ldots, y^n; \lambda_1, \ldots, \lambda_m)$ be a function of $1 + n + m$ variables $t, y^1, \ldots, y^n$, $\lambda_1, \ldots, \lambda_m$. Assume that these functions have up to the k-th partial derivatives on a domain $|t| < a$, $|y^i| < b_i$, $|\lambda_j| < c_j$ which are continuous. Then if we consider a system of differential equations*

$$\frac{dy^i}{dt} = f^i(t; y^1, \ldots, y^n, \lambda_1, \ldots, \lambda_m) \qquad (i = 1, \ldots, n)$$

with respect to unknown functions $y^i = y^i(t; \lambda_1, \ldots, \lambda_m)$ $(i = 1, \ldots, n)$, there exists a unique set of solutions $y^i(t; \lambda_1, \ldots, \lambda_m)$ such that they satisfy the initial condition

$$y^i(\alpha; \lambda_1, \ldots, \lambda_m) = \beta^i \qquad (i = 1, \ldots, n).$$

The solutions are defined on $|t - a| < \delta$ for a sufficiently small $\delta > 0$, $k + 1$ times continuously differentiable on t (i.e., of C^{k+1} class on t), and k times continuously differentiable on parameters $\lambda_1, \ldots, \lambda_m$ (i.e., of C^k class on $\lambda_1, \ldots, \lambda_m$).

The equation of a geodesic is

$$\frac{d^2u^i}{dt^2} + \sum_{j,k=1}^{2} \Gamma^i_{jk}(u^1, u^2) \frac{du^j}{dt} \frac{du^k}{dt} = 0 \qquad (i = 1, 2),$$

which is the second order differential equation. If we set

$$y^i = \frac{du^i}{dt} \qquad (i = 1, 2),$$

$$y^{2+i} = u^i \qquad (i = 1, 2),$$

S. Kobayashi, *Differential Geometry of Curves and Surfaces*,
Springer Undergraduate Mathematics Series,
https://doi.org/10.1007/978-981-15-1739-6

the equation reduces the following system of the first order differential equations:

$$\frac{dy^i}{dt} = -\sum_{j,k=1}^{2} \Gamma^i_{jk}(y^3, y^4)\, y^j\, y^k \qquad (i = 1, 2),$$

$$\frac{dy^{2+i}}{dt} = y^i \qquad (i = 1, 2).$$

The initial condition

$$u^i(0) = \alpha^i, \qquad \frac{du^i}{dt}(0) = \beta^i$$

is changed to

$$y^i(0) = \beta^i, \qquad y^{2+i}(0) = \alpha^i$$

and we can apply the above theorem.

Solutions to Problems

Chapter 1

Problem 1.2.1 If we set $\boldsymbol{p}(t) = (u(t), v(t))$, we get

$$\boldsymbol{e}_1 = \frac{d\boldsymbol{p}}{ds} = \frac{d\boldsymbol{p}}{dt}\frac{dt}{ds} = \left(\dot{x}(t)\frac{dt}{ds}, \dot{y}(t)\frac{dt}{ds}\right)$$

as (1.2.44). On the other hand, we have

$$\frac{dt}{ds} = \frac{1}{\sqrt{\dot{x}(t)^2 + \dot{y}(t)^2}}$$

by (1.2.42). Hence

$$\boldsymbol{e}_1 = \left(\frac{\dot{x}(t)}{\sqrt{\dot{x}(t)^2 + \dot{y}(t)^2}}, \frac{\dot{y}(t)}{\sqrt{\dot{x}(t)^2 + \dot{y}(t)^2}}\right).$$

Rotating $\boldsymbol{e}_1$ by $90°$, we get

$$\boldsymbol{e}_2 = \left(\frac{-\dot{y}(t)}{\sqrt{\dot{x}(t)^2 + \dot{y}(t)^2}}, \frac{\dot{x}(t)}{\sqrt{\dot{x}(t)^2 + \dot{y}(t)^2}}\right).$$

Next, (1.2.47) follows

$$\begin{aligned}\frac{d\boldsymbol{e}_1}{ds} = \frac{d\boldsymbol{e}_1}{dt}\frac{dt}{ds} &= \left(\frac{-(\dot{x}\ddot{y} - \ddot{x}\dot{y})\dot{y}}{(\dot{x}^2 + \dot{y}^2)^{\frac{3}{2}}}, \frac{(\dot{x}\ddot{y} - \ddot{x}\dot{y})\dot{x}}{(\dot{x}^2 + \dot{y}^2)^{\frac{3}{2}}}\right)\frac{1}{\sqrt{\dot{x}^2 + \dot{y}^2}} \\ &= \frac{\dot{x}\ddot{y} - \ddot{x}\dot{y}}{(\dot{x}^2 + \dot{y}^2)^{\frac{3}{2}}}\left(\frac{-\dot{y}}{\sqrt{\dot{x}^2 + \dot{y}^2}}, \frac{\dot{x}}{\sqrt{\dot{x}^2 + \dot{y}^2}}\right) = \frac{\dot{x}\ddot{y} - \ddot{x}\dot{y}}{(\dot{x}^2 + \dot{y}^2)^{\frac{3}{2}}}\boldsymbol{e}_2.\end{aligned}$$

Thus by (1.2.21)

$$\kappa = \frac{\dot{x}\ddot{y} - \ddot{x}\dot{y}}{(\dot{x}^2 + \dot{y}^2)^{\frac{3}{2}}}.$$

S. Kobayashi, *Differential Geometry of Curves and Surfaces*,
Springer Undergraduate Mathematics Series,
https://doi.org/10.1007/978-981-15-1739-6

Problem 1.2.2 Substituting $r = F(\theta)$ in the relation between the polar coordinates and the usual coordinates x, y:

$$x = r\cos\theta, \qquad y = r\sin\theta,$$

we consider x, y as functions of θ. Setting $t = \theta$, we use (1.2.50) proved in Problem 1.2.1. Substituting

$$\begin{aligned}
\frac{dx}{d\theta} &= \frac{dr}{d\theta}\cos\theta - r\sin\theta, \qquad \frac{dy}{d\theta} = \frac{dr}{d\theta}\sin\theta + r\cos\theta,\\
\frac{d^2x}{d\theta^2} &= \frac{d^2r}{d\theta^2}\cos\theta - 2\frac{dr}{d\theta}\sin\theta - r\cos\theta,\\
\frac{d^2y}{d\theta^2} &= \frac{d^2r}{d\theta^2}\sin\theta + 2\frac{dr}{d\theta}\cos\theta - r\sin\theta
\end{aligned}$$

for (1.2.50) and using $\sin^2\theta + \cos^2\theta = 1$ to simplify it, we obtain the equation to be proved.

Problem 1.2.3 We have

$$\begin{aligned}
x^2 - y^2 = \cosh^2 t - \sinh^2 t &= \left(\frac{e^t + e^{-t}}{2}\right)^2 - \left(\frac{e^t - e^{-t}}{2}\right)^2\\
&= \frac{e^{2t} + 2 + e^{-2t}}{4} - \frac{e^{2t} - 2 + e^{-2t}}{4} = 1.
\end{aligned}$$

Next we calculate κ by using (1.2.50) proved in Problem 1.2.1. The definition of $\cosh t$ and $\sinh t$ implies

$$\begin{aligned}
\dot{x} &= \sinh t, \qquad \ddot{x} = \cosh t,\\
\dot{y} &= \cosh t, \qquad \ddot{y} = \sinh t.
\end{aligned}$$

Thus we have

$$\kappa = \frac{\sinh^2 t - \cosh^2 t}{(\sinh^2 t + \cosh^2 t)^{\frac{3}{2}}} = \frac{-1}{\left(\dfrac{e^{2t} + e^{-2t}}{2}\right)^{\frac{3}{2}}} = \frac{-1}{(\cosh 2t)^{\frac{3}{2}}}.$$

Here the curvature κ is negative. In some books $\boldsymbol{e}_2$ is chosen so that κ is positive. Since $\cosh t$ is always positive, we can only obtain the right half of the hyperbola by this parameter representation. (We can obtain the left half of the hyperbola if we set $x = -\cosh t$.) When t moves from $-\infty$ to ∞, (x, y) moves from the lower one to the upper one. The fact that κ approaches 0 when t goes to $\pm\infty$ in the above equation of κ indicates that the hyperbola does not curve around $t = \pm\infty$, that is, it is almost a straight line. In fact, it approaches asymptotes $y = \pm x$.

Problem 1.2.4 The arc length is obtained from

$$s = \int_0^x \sqrt{1 + \left(\frac{dy}{dx}\right)^2}\,dx = \int_0^x \sqrt{1 + \sinh^2\frac{x}{a}}\,dx = \int_0^x \cosh\frac{x}{a}\,dx = a\sinh\frac{x}{a}.$$

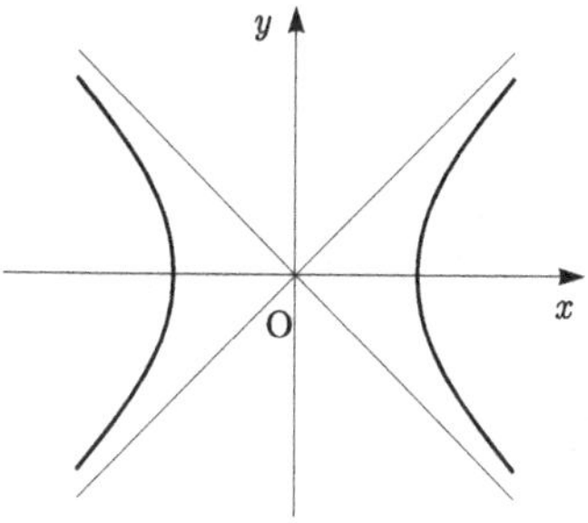

Fig. A.1

Hence,

$$x = a \sinh^{-1} \frac{s}{a},$$

$$y = a \cosh \frac{x}{a} = a\sqrt{1 + \sinh^2 \frac{x}{a}} = a\sqrt{1 + \frac{s^2}{a^2}} = \sqrt{a^2 + s^2}.$$

The parameter representation we seek is

$$x = a \sinh^{-1} \frac{s}{a}, \qquad y = \sqrt{a^2 + s^2}.$$

The name "catenary" is originated from the fact that if we hang a string with uniform density in a gravitational field keeping both ends, the shape of the string is a curve which is in the shape of the graph of a catenary. The Latin word "catena" is the original word for "chain."

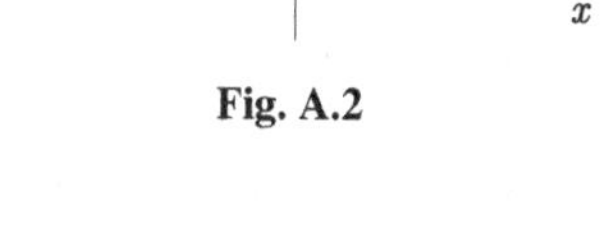

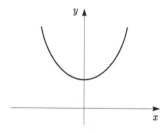

Fig. A.2

Problem 1.3.1 (a) The meaning of $\boldsymbol{p}(t) \cdot \boldsymbol{e}_2(t)$ is that when we project a vector $\boldsymbol{p}(t)$ orthogonally onto $\boldsymbol{e}_2(t)$, it is $\boldsymbol{p}(t) \cdot \boldsymbol{e}_2(t)$ times $\boldsymbol{e}_2(t)$. The meaning of $\boldsymbol{p}(t+\pi) \cdot \boldsymbol{e}_2(t+\pi)$ is similar. Other things are clear from Figure A.3.

(b) $\int_0^\pi W(t)dt$

$$= \int_0^\pi -\{\boldsymbol{p}(t) \cdot \boldsymbol{e}_2(t) + \boldsymbol{p}(t+\pi) \cdot \boldsymbol{e}_2(t+\pi)\}\, dt = \int_0^\pi -\boldsymbol{p}(t) \cdot \boldsymbol{e}_2(t)\, dt.$$

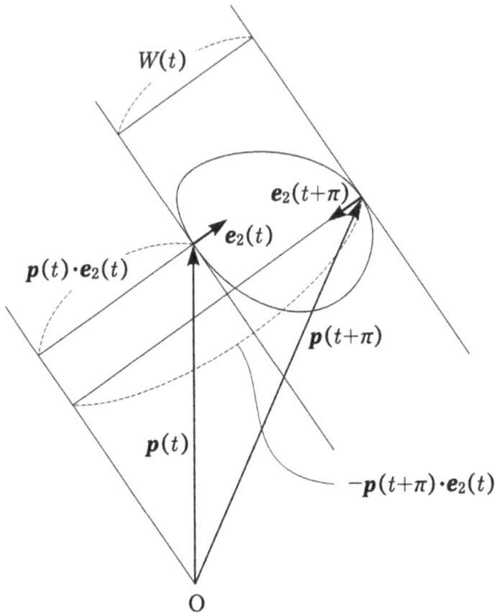

Fig. A.3

On the other hand, since

$$\frac{d\boldsymbol{e}_1}{dt} = \frac{d\boldsymbol{e}_1}{ds}\frac{dt}{ds} = \kappa \boldsymbol{e}_2 \frac{ds}{dt} = \boldsymbol{e}_2,$$

we have

$$\begin{aligned}\int_0^{\pi} W(t)dt &= \int_0^{2\pi} -\boldsymbol{p}\cdot\frac{d\boldsymbol{e}_1}{dt}\,dt = \int_0^{2\pi}\left\{-\frac{d}{dt}(\boldsymbol{p}\cdot\boldsymbol{e}_1) + \frac{d\boldsymbol{p}}{dt}\cdot\boldsymbol{e}_1\right\}dt \\ &= -[\boldsymbol{p}\cdot\boldsymbol{e}_1]_0^{2\pi} + \int_0^{2\pi}\frac{d\boldsymbol{p}}{dt}\cdot\boldsymbol{e}_1\,dt = \int_0^{2\pi}\frac{d\boldsymbol{p}}{dt}\cdot\boldsymbol{e}_1\frac{ds}{dt}\,dt \\ &= \int_0^{2\pi}\boldsymbol{e}_1\cdot\boldsymbol{e}_1\frac{ds}{dt}\,dt = \int_0^{2\pi}\frac{ds}{dt}\,dt = L.\end{aligned}$$

A circle is the simplest example of curves with constant width and the width is equal to its diameter. The formula $L = \pi W$ is nothing but the formula of the circumference. Another example of a curve with constant width is the Reuleaux's triangle. First draw an equilateral triangle, then draw three circles where their radius are the length of each side of the equilateral triangle centered at each vertex. Then three shorter arcs make an oval which is not smooth at each vertex. This is the Reuleaux's triangle and if a circle whose center lies on the Reuleaux's triangle moves along this triangle with constant radius, the outside curve is a smooth curve with constant width. This is called a curve parallel to the Reuleaux's triangle (see Figure A.4).

By the way, the shape of a manhole cover is not a square or a triangle but a circle. This is to avoid accidentally dropping the manhole cover into the hole when we lift

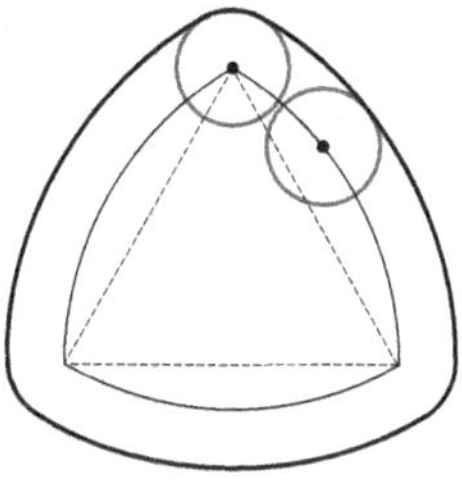

Fig. A.4

it up. Actually, it does not have to be a circle. It could be any curve with constant width.

Problem 1.3.2 If we write $z = re^{i\theta}$ by using the polar coordinate, a curve $|z| = 1$ is given as $z = e^{i\theta}$ with parameter θ. If we write $w = u + iv$, the curvature $\kappa(\theta)$ is

$$\kappa(\theta) = \frac{\dfrac{du}{d\theta}\dfrac{d^2v}{d\theta^2} - \dfrac{d^2u}{d\theta^2}\dfrac{dv}{d\theta}}{\left\{\left(\dfrac{du}{d\theta}\right)^2 + \left(\dfrac{dv}{d\theta}\right)^2\right\}^{\frac{3}{2}}}$$

by (1.2.50). If we use

$$\frac{d^2w}{d\theta^2}\frac{d\bar{w}}{d\theta} = \left(\frac{d^2u}{d\theta^2}\frac{du}{d\theta} + \frac{d^2v}{d\theta^2}\frac{dv}{d\theta}\right) + i\left(\frac{du}{d\theta}\frac{d^2v}{d\theta^2} - \frac{d^2u}{d\theta^2}\frac{dv}{d\theta}\right),$$

the numerator of the above equation of $\kappa(\theta)$ is the imaginary part of $\frac{d^2w}{d\theta^2}\frac{d\bar{w}}{d\theta}$ and so $\mathrm{Im}\left[\frac{d^2w}{d\theta^2}\frac{d\bar{w}}{d\theta}\right]$. The denominator is $\left[\frac{dw}{d\theta}\frac{d\bar{w}}{d\theta}\right]^{\frac{3}{2}}$. Moreover, $\frac{dz}{d\theta} = ie^{i\theta}$ implies

$$\frac{dw}{d\theta} = \frac{dw}{dz}\frac{dz}{d\theta} = f'(z)\,ie^{i\theta}, \qquad \frac{d\bar{w}}{d\theta} = -\bar{f}'(z)\,ie^{i\theta},$$
$$\frac{d^2w}{d\theta^2} = -f''(z)\,e^{2i\theta} - f'(z)\,e^{i\theta}.$$

Hence the curvature is

$$\kappa(\theta) = \frac{\mathrm{Re}(f''\,\bar{f}'\,e^{i\theta} + f'\,\bar{f}')}{(f'\,\bar{f}')^{\frac{3}{2}}} \qquad \text{(Re means the real part).}$$

In order to calculate the rotation index we write

$$ds = \frac{ds}{d\theta}\,d\theta = \left|\frac{df}{d\theta}\right|\,d\theta = \left|\frac{df}{dz}\right|\,d\theta,$$

and by (2.3.5)

$$m = \frac{1}{2\pi}\int_0^{2\pi} \kappa \cdot |f'|\, d\theta = \frac{1}{2\pi}\int_0^{2\pi} \frac{\mathrm{Re}(f''\bar{f}' e^{i\theta} + f'\bar{f}')}{f'\bar{f}'}\, d\theta$$
$$= \frac{1}{2\pi}\int_0^{2\pi}\left(\mathrm{Re}\frac{f'' e^{i\theta}}{f'} + 1\right) d\theta = \mathrm{Re}\left[\frac{1}{2\pi}\int_0^{2\pi} \frac{f'' e^{i\theta}}{f'}\, d\theta\right] + 1$$
$$= \mathrm{Re}\left[\frac{1}{2\pi i}\oint_{|z|=1} \frac{f''}{f'}\, dz\right] + 1.$$

According to the argument principle of complex function theory, the integral

$$\frac{1}{2\pi i}\oint_{|z|=1} \frac{f''}{f'}\, dz$$

expresses the number of zeros of f' in $|z| < 1$. Thus m is obtained by adding 1 to the number of zeros of f' in $|z| < 1$.

Problem 1.4.1 We should substitute

$$\boldsymbol{p}' = \boldsymbol{e}_1, \quad \boldsymbol{p}'' = \boldsymbol{e}_1' = \kappa \boldsymbol{e}_2, \quad \boldsymbol{p}''' = (\kappa \boldsymbol{e}_2)' = \kappa' \boldsymbol{e}_2 - \kappa^2 \boldsymbol{e}_1 + \kappa\tau \boldsymbol{e}_3$$

into the Taylor expansion of $\boldsymbol{p}(s)$:

$$\boldsymbol{p}(s) = \boldsymbol{p}(0) + \boldsymbol{p}'(0)s + \boldsymbol{p}''(0)\frac{s^2}{2!} + \boldsymbol{p}'''(0)\frac{s^3}{3!} + \cdots .$$

Problem 1.4.2

$$\dot{\boldsymbol{p}} = \boldsymbol{p}'\frac{ds}{dt} = \frac{ds}{dt}\boldsymbol{e}_1, \quad \ddot{\boldsymbol{p}} = \kappa\left(\frac{ds}{dt}\right)^2 \boldsymbol{e}_2 + \frac{d^2 s}{dt^2}\boldsymbol{e}_1$$
$$\dddot{\boldsymbol{p}} = \kappa\tau\left(\frac{ds}{dt}\right)^3 \boldsymbol{e}_3 + \cdots \quad (\cdots \text{ means terms of } \boldsymbol{e}_1 \text{ and } \boldsymbol{e}_2).$$

Therefore

$$\dot{\boldsymbol{p}} \times \ddot{\boldsymbol{p}} = \kappa\left(\frac{ds}{dt}\right)^3 \boldsymbol{e}_3, \quad |\dot{\boldsymbol{p}} \times \ddot{\boldsymbol{p}}| = \kappa\left|\frac{ds}{dt}\right|^3, \quad |\dot{\boldsymbol{p}}| = \left|\frac{ds}{dt}\right|,$$
$$|\dot{\boldsymbol{p}}\ \ddot{\boldsymbol{p}}\ \dddot{\boldsymbol{p}}| = \left|\frac{ds}{dt}\boldsymbol{e}_1\ \ \kappa\left(\frac{ds}{dt}\right)^2\boldsymbol{e}_2\ \ \kappa\tau\left(\frac{ds}{dt}\right)^3\boldsymbol{e}_3\right| = \kappa^2\tau\left(\frac{ds}{dt}\right)^6.$$

We obtain (1.4.41) easily by these equations.

Problem 1.4.3 Since

$$\boldsymbol{p}' = \boldsymbol{e}_1, \quad 0 = (\boldsymbol{e}_1 \cdot \boldsymbol{e}_1)' = 2\boldsymbol{e}_1 \cdot \boldsymbol{e}_1',$$

we define

$$\boldsymbol{e}_1' = \kappa_1 \boldsymbol{e}_2 \quad (\kappa_1 > 0).$$

We need to assume $\kappa_1 = |\boldsymbol{e}_1'| > 0$ in order to define $\boldsymbol{e}_2$. Next, since we have

$$0 = (\boldsymbol{e}_2 \cdot \boldsymbol{e}_2)' = 2\boldsymbol{e}_2 \cdot \boldsymbol{e}_2',$$
$$0 = (\boldsymbol{e}_1 \cdot \boldsymbol{e}_2)' = \boldsymbol{e}_1' \cdot \boldsymbol{e}_2 + \boldsymbol{e}_1 \cdot \boldsymbol{e}_2' = \kappa_1 + \boldsymbol{e}_1 \cdot \boldsymbol{e}_2',$$

if $\kappa_2 = |e_2' + \kappa_1 e_1| > 0$, we can define e_3 as

$$e_2' = -\kappa_1 e_1 + \kappa_2 e_3 \quad \text{i.e.,} \ e_3 = \frac{1}{\kappa_2}(e_2' + \kappa_1 e_1).$$

Chapter 2

Problem 2.1.1 (i) Example 2.1.5, elliptic paraboloid. (ii) Example 2.1.6, hyperbolic paraboloid. (iii) Example 2.1.6, hyperbolic paraboloid. (iv) Example 2.1.3, hyperboloid of one sheet.

Problem 2.2.1 Since $p(x, y) = (x, y, f(x, y))$, we have

$$\begin{aligned} p_x &= (1,0,p), \quad p_y = (0,1,q), \\ p_x \times p_y &= \left(\begin{vmatrix} 0 & p \\ 1 & q \end{vmatrix}, -\begin{vmatrix} 1 & p \\ 0 & q \end{vmatrix}, \begin{vmatrix} 1 & 0 \\ 0 & 1 \end{vmatrix} \right) = (-p, -q, 1), \\ e &= \frac{p_x \times p_y}{|p_x \times p_y|} = \left(\frac{-p}{\sqrt{1+p^2+q^2}}, \frac{-q}{\sqrt{1+p^2+q^2}}, \frac{1}{\sqrt{1+p^2+q^2}} \right), \end{aligned}$$

$$\begin{aligned} \mathbf{I} = dp \cdot dp &= (dx, dy, pdx + qdy) \cdot (dx, dy, pdx + qdy) \\ &= dxdx + dydy + p^2\, dxdx + 2pq\, dxdy + q^2\, dydy \\ &= (1+p^2)\, dxdx + 2pq\, dxdy + (1+q^2)\, dydy. \end{aligned}$$

Next, we will calculate $\mathbf{II}$. To avoid calculating de, we will use (2.2.23) instead of (2.2.20).Thus

$$\begin{aligned} L &= (0,0,r) \cdot e = \frac{r}{\sqrt{1+p^2+q^2}}, \qquad M = (0,0,s) \cdot e = \frac{s}{\sqrt{1+p^2+q^2}}, \\ N &= (0,0,t) \cdot e = \frac{t}{\sqrt{1+p^2+q^2}}. \end{aligned}$$

Lastly, we can calculate K and H easily by (2.2.43).

Problem 2.2.2 Differentiating $e \cdot e = 1$, we obtain $e \cdot e_u = 0$. Therefore we can write $e_u = A p_u + B p_v$. We get

$$\begin{aligned} -L &= e_u \cdot p_u = AE + BF, \\ -M &= e_u \cdot p_v = AF + BG \end{aligned}$$

by (2.2.23). Solving them with respect to A and B, we obtain

$$A = \frac{MF - LG}{EG - F^2}, \qquad B = \frac{LF - ME}{EG - F^2}.$$

Hence

$$e_u = \frac{MF - LG}{EG - F^2} p_u + \frac{LF - ME}{EG - F^2} p_v.$$

Similarly,

$$e_v = \frac{NF - MG}{EG - F^2} p_u + \frac{MF - NE}{EG - F^2} p_v.$$

Substituting this into $\mathbf{III} = de \cdot de = (e_u du + e_v dv) \cdot (e_u du + e_v dv)$, we obtain

$$\mathbf{III} = \frac{L^2G + M^2E - 2LMF}{EG - F^2} dudu + 2\frac{LMG - M^2F - LNF + MNE}{EG - F^2} dudv + \frac{M^2G + N^2E - 2MNF}{EG - F^2} dvdv.$$

By using (2.2.43), we can write

$$\begin{aligned} \mathbf{III} &= (2HL - KE)dudu + 2(2HM - KF)dudv + (2HN - KG)dvdv \\ &= 2H(L\,dudu + 2M\,dudv + N\,dvdv) - K(E\,dudu + 2F\,dvdv + G\,dvdv) \\ &= 2H\mathbf{II} - K\mathbf{I}. \end{aligned}$$

Problem 2.2.3 Substitute (2.2.25) into

$$p'(s) = p_u \frac{du}{ds} + p_v \frac{dv}{ds}$$

and

$$p''(s) = p_u \frac{d^2u}{ds^2} + p_{uu} \frac{du}{ds}\frac{du}{ds} + 2p_{uv} \frac{du}{ds}\frac{dv}{ds} + p_{vv} \frac{dv}{ds}\frac{dv}{ds} + p_v \frac{d^2v}{ds^2},$$

to express it as a linear combination of p_u, p_v, e and to set the coefficients of p_u and p_v to 0.
Problem 2.2.4 By $d\bar{p} = dp + \varepsilon f de + \varepsilon df \cdot e$ and $e \cdot de = e \cdot dp = 0$ we get

$$\begin{aligned} d\bar{p} \cdot d\bar{p} &= dp \cdot dp + 2\varepsilon f dp \cdot de + \varepsilon^2(\cdots) \\ &= (E - 2\varepsilon f L)dudu + 2(F - 2\varepsilon f M)dudv + (G - 2\varepsilon f N)dvdu + \varepsilon^2(\cdots). \end{aligned}$$

Hence

$$\begin{aligned} A(\varepsilon) &= \iint_R \sqrt{(E - 2\varepsilon f L)(G - 2\varepsilon f N) - (F - 2\varepsilon f M)^2 + \varepsilon^2(\cdots)}\, dudv \\ &= \iint_R \sqrt{EG - F^2 - 2\varepsilon f(EN + LG - 2FM) + \varepsilon^2(\cdots)}\, dudv. \end{aligned}$$

Differentiating it with respect to ε and putting $\varepsilon = 0$, we obtain

$$\begin{aligned} A'(0) &= \iint_R \frac{-f(EN + LG - 2FM)}{\sqrt{EG - F^2}} dudv \\ &= -\iint_R fH\sqrt{EG - F^2}\, dudv. \end{aligned}$$

Problem 2.2.5 It is sufficient to prove that the unit vector e is constant. Since $e_u \cdot p_u = -L = 0$, $e_u \cdot p_v = -M = 0$, and $e_u \cdot e = \left(\frac{1}{2} e \cdot e\right)_u = 0$, we have $e_u = \mathbf{0}$. Similarly, $e_v = \mathbf{0}$. Therefore e is constant.
Problem 2.2.6 (a) Since

$$\boldsymbol{p}_u = \boldsymbol{q}'(u) + v\boldsymbol{t}'(u), \quad \boldsymbol{p}_v = \boldsymbol{t}(u),$$
$$\boldsymbol{p}_{uu} = \boldsymbol{q}''(u) + v\boldsymbol{t}''(u), \quad \boldsymbol{p}_{uv} = \boldsymbol{p}_{vu} = \boldsymbol{t}'(u), \quad \boldsymbol{p}_{vv} = \boldsymbol{0},$$

we get $N = \boldsymbol{p}_{vv} \cdot \boldsymbol{e} = 0$. Thus $K = -\frac{M^2}{EG-F^2} \leq 0$.

(b) By the proof of (a) we know $K = 0$ is equivalent to $M = 0$. Since $M = \boldsymbol{p}_{uv} \cdot \boldsymbol{e} = \boldsymbol{t}' \cdot \boldsymbol{e}$, $M = 0$ if and only if $\boldsymbol{t}'$ is tangent to the surface, that is, $\boldsymbol{t}'$ is expressed as a linear combination of $\boldsymbol{p}_u$ and $\boldsymbol{p}_v$. On the other hand, $\boldsymbol{p}_u = \boldsymbol{q}' + v\boldsymbol{t}'$ and $\boldsymbol{p}_v = \boldsymbol{t}$. Therefore $\boldsymbol{t}'$ is expressed as a linear combination of $\boldsymbol{p}_u$ and $\boldsymbol{p}_v$ means nothing but $\boldsymbol{t}'$ is expressed as a linear combination of $\boldsymbol{q}'$ and $\boldsymbol{t}$.

Now we only need to show that the normal direction of the tangent plane is independent from v. The direction of the normal vector $\boldsymbol{p}_u \times \boldsymbol{p}_v = (\boldsymbol{q}'(u) + v\boldsymbol{t}'(u)) \times \boldsymbol{t}(u)$ is independent from v if and only if $\boldsymbol{t}'(u) \times \boldsymbol{t}(u)$ is proportional to $\boldsymbol{q}'(u) \times \boldsymbol{t}(u)$, that is, $\boldsymbol{t}'(u)$ is a linear combination of $\boldsymbol{q}'(u)$ and $\boldsymbol{t}(u)$.

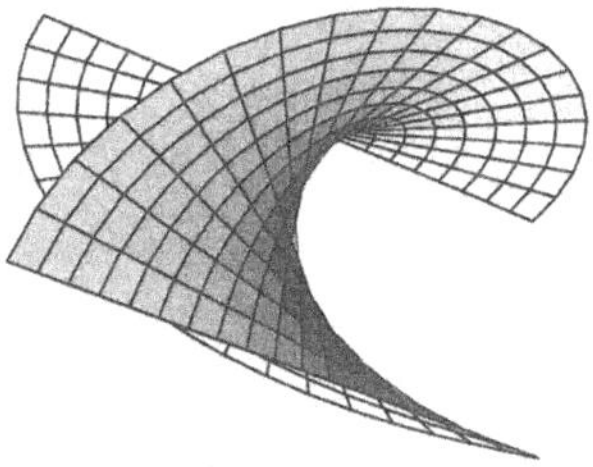

Fig. A.5

Problem 2.2.7 (a) The simplest way to express the curve $\boldsymbol{q}(u)$ is to use the ellipse which is the intersection of the surface and the (x, y)-plane. Thus

$$\boldsymbol{q}(u) = (a \cos u,\ b \sin u,\ 0).$$

Here we note that u is not the arc length parameter. Next, we set $\boldsymbol{t}(u) = (\xi(u), \eta(u), \zeta(u))$ and determine $\boldsymbol{t}(u)$ so that $\boldsymbol{q}(u) + v\boldsymbol{t}(u)$ satisfies the equation of the hyperboloid of one sheet. Substituting

$$x = a \cos u + v\xi(u), \quad y = b \sin u + v\eta(u), \quad z = v\zeta(u)$$

into the equation of the hyperboloid of one sheet, we obtain

$$\frac{(a \cos u + v\xi(u))^2}{a^2} + \frac{(b \sin u + v\eta(u))^2}{b^2} - \frac{(v\zeta(u))^2}{c^2} = 1.$$

By calculating this equation, we get

$$2v\left(\frac{\xi(u)\cos u}{a} + \frac{\eta(u)\sin u}{b}\right) + v^2\left(\frac{\xi(u)^2}{a} + \frac{\eta(u)^2}{b} - \frac{\zeta(u)^2}{c^2}\right) = 0.$$

Since the above equation needs to hold for all v, we have to determine ξ, η, ζ so that both coefficients of $2v$ and v^2 are 0. Therefore it turns out that we should set

$$\xi(u) = -a\sin u, \quad \eta(u) = b\cos u, \quad \zeta(u) = \pm c.$$

According to $\zeta(u) = \pm c$, we find that a hyperboloid of one sheet is a ruled surface in two ways (see Figure A.6).

(b) Let $\boldsymbol{q}(u) = \left(u,\ 0,\ -\frac{u^2}{a^2}\right)$ be the parabola which is the intersection of the surface and the (x, z)-plane. With this $\boldsymbol{q}(u)$, with a calculation similar as the one before, we obtain

$$\boldsymbol{t}(u) = \left(a,\ \pm b,\ -\frac{2u}{a}\right).$$

A hyperbolic paraboloid is also a ruled surface in two ways (see Figure A.7).

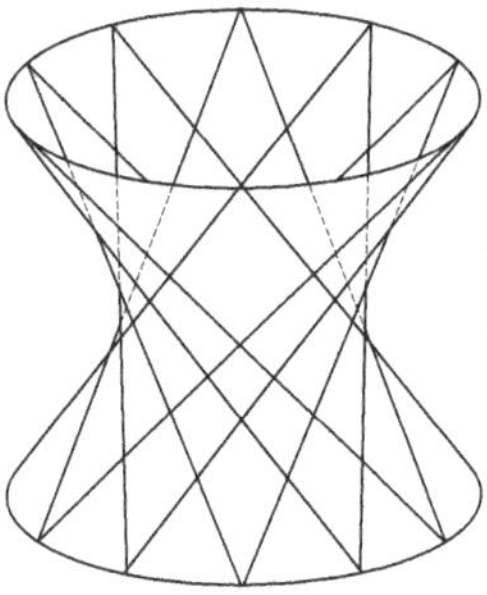

Fig. A.6

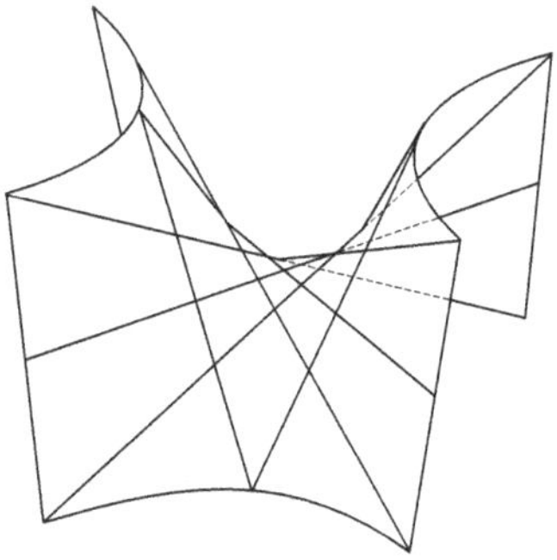

Fig. A.7

Problem 2.3.1 (a) In the case of an elliptic paraboloid $z = \frac{x^2}{a^2} + \frac{y^2}{b^2}$, since it is a surface in the form of Problem 2.2.1, by using notations in Problem 2.2.1 we get

$$p = \frac{2x}{a^2}, \quad q = \frac{2y}{b^2}, \quad r = \frac{2}{a^2}, \quad s = 0, \quad t = \frac{2}{b^2}.$$

Hence

$$K = \frac{4}{a^2b^2\left(1 + \frac{4x^2}{a^4} + \frac{4y^2}{b^4}\right)^2},$$

$$H = \frac{\frac{2}{a^2}\left(1 + \frac{4y^2}{b^4}\right) + \frac{2}{b^2}\left(1 + \frac{4x^2}{a^4}\right)}{2\left(1 + \frac{4x^2}{a^4} + \frac{4y^2}{b^4}\right)^{\frac{3}{2}}}.$$

(a') When an elliptic paraboloid is expressed as $x = au\cos v$, $y = bu\sin v$, $z = u^2$.

$$\boldsymbol{p}_u = (a\cos v,\ b\sin v,\ 2u), \quad \boldsymbol{p}_v = (-au\sin v,\ bu\cos v,\ 0),$$
$$\boldsymbol{p}_{uu} = (0,\ 0,\ 2), \quad \boldsymbol{p}_{uv} = \boldsymbol{p}_{vu} = (-a\sin v, b\cos v, 0),$$
$$\boldsymbol{p}_{vv} = (-au\cos v,\ -bu\sin v,\ 0),$$
$$\boldsymbol{e} = \frac{1}{\Delta}(-2bu\cos v,\ -2au\sin v,\ ab), \quad \text{where } \Delta = \sqrt{a^2b^2 + 4a^2u^2\sin^2 v + 4b^2u^2\cos^2 v},$$
$$E = a^2\cos^2 v + b^2\sin^2 v + 4u^2, \quad F = (b^2 - a^2)u\sin v\cos v,$$
$$G = a^2u^2\sin 2v + b^2u^2\cos^2 v, \quad EG - F^2 = u^2\Delta^2,$$
$$L = \frac{2ab}{\Delta}, \quad M = 0, \quad N = \frac{2abu^2}{\Delta},$$
$$K = \frac{4a^2b^2}{\Delta^4}, \quad H = \frac{ad(a^2 + b^2 + 4u^2)}{\Delta^3}.$$

(b) The case where the hyperbolic paraboloid is expressed as $z = \frac{x^2}{a^2} - \frac{y^2}{b^2}$.

$$p = \frac{2x}{a^2}, \quad q = -\frac{2y}{b^2}, \quad r = \frac{2}{a^2}, \quad s = 0, \quad t = -\frac{2}{b^2},$$
$$K = \frac{-4}{a^2b^2\left(1 + \frac{4x^2}{a^4} + \frac{4y^2}{b^4}\right)^2},$$
$$H = \frac{\frac{2}{a^2}\left(1 + \frac{4y^2}{b^4}\right) - \frac{2}{b^2}\left(1 + \frac{4x^2}{a^4}\right)}{2\left(1 + \frac{4x^2}{a^4} + \frac{4y^2}{b^4}\right)^{\frac{3}{2}}}.$$

(b') The case where we express an hyperbolic paraboloid as $x = au\cosh v,\ y = bu\sinh v,\ z = u^2$.

$$\boldsymbol{p}_u = (a\cosh v,\ b\sinh v,\ 2u), \quad \boldsymbol{p}_v = (au\sinh v,\ bu\cosh v,\ 0),$$
$$\boldsymbol{p}_{uu} = (0,\ 0,\ 2), \quad \boldsymbol{p}_{uv} = \boldsymbol{p}_{vu} = (a\sinh v,\ b\cosh v,\ 0),$$
$$\boldsymbol{p}_{vv} = (au\cosh v,\ bu\sinh v,\ 0),$$
$$\boldsymbol{e} = \frac{1}{\Delta}(-2bu\cosh v,\ 2au\sinh v,\ ab), \quad \text{where } \Delta = \sqrt{a^2b^2 + 4a^2u^2\sinh^2 v + 4b^2u^2\cosh^2 v},$$
$$E = a^2\cosh^2 v; b^2\sinh^2 v + 4u^2, \quad F = (a^2 + b^2)u\cosh v\sinh v,$$
$$G = a^2u^2\sinh^2 v + b^2u^2\cosh^2 v, \quad EG - F^2 = u^2\Delta^2,$$
$$L = \frac{2ab}{\Delta}, \quad M = 0, \quad N = \frac{-2abu^2}{\Delta},$$
$$K = \frac{-4a^2b^2}{\Delta^4}, \quad H = \frac{ab(-a^2 + b^2 - 4u^2)}{\Delta^3}.$$

(b'') The case where we express an hyperbolic paraboloid as $x = a(u + v),\ y = b(u - v),\ z = 4uv$.

$$\begin{aligned}
&\boldsymbol{p}_u = (a, b, 4v), \quad \boldsymbol{p}_v = (a, -b, 4u), \\
&\boldsymbol{p}_{uu} = (0, 0, 0), \quad \boldsymbol{p}_{uv} = \boldsymbol{p}_{vu} = (0, 0, 4), \quad \boldsymbol{p}_{vv} = (0, 0, 0), \\
&\boldsymbol{e} = \frac{1}{\Delta}(2b(u+v), -2a(u-v), -ab), \quad \text{where } \Delta = \sqrt{4a^2(u-v)^2 + 4b^2(u+v)^2 + a^2b^2}, \\
&E = a^2 + b^2 + 16v^2, \quad F = a^2 - b^2 + 16uv, \\
&G = a^2 + b^2 + 16u^2, \quad EG - F^2 = 4\Delta^2, \\
&L = 0, \quad M = -\frac{4ab}{\Delta}, \quad N = 0, \\
&K = \frac{-4a^2b^2}{\Delta^4}, \quad H = \frac{ab(a^2 - b^2 + 16uv)}{\Delta^3}.
\end{aligned}$$

Problem 2.3.2

$$\begin{aligned}
&\boldsymbol{p}_u = (\cos v, \sin v, 0), \quad \boldsymbol{p}_v = (-u \sin v, u \cos v, f'(v)) \\
&\boldsymbol{p}_{uu} = (0, 0, 0), \quad \boldsymbol{p}_{uv} = \boldsymbol{p}_{vu} = (-\sin v, \cos v, 0), \\
&\boldsymbol{p}_{vv} = (-u \cos v, -u \sin v, f''(v)) \\
&\boldsymbol{e} = \frac{1}{\Delta}(f'(v) \sin v, -f'(v) \cos v, u), \quad \text{where } \Delta = \sqrt{u^2 + f'(v)^2}, \\
&E = 1, \quad F = 0, \quad G = u^2 + f'(v)^2, \quad EG - F^2 = \Delta^2, \\
&L = 0, \quad M = -\frac{f'(v)}{\Delta}, \quad N = \frac{uf''(v)}{\Delta}, \\
&K = -\frac{f'(v)^2}{(u^2 + f'(v)^2)^2}, \quad H = \frac{uf''(v)}{2(u^2 + f'(v)^2)^{\frac{3}{2}}}.
\end{aligned}$$

If $f(v) = av + b$,

$$K = -\frac{a^2}{(u^2 + a^2)^2}, \quad H = 0.$$

Conversely, if $H = 0$, we get $f''(v) = 0$. Thus $f(v) = av + b$.

Problem 2.3.3

$$\begin{aligned}
&\boldsymbol{p}_u = (3 + 3v^2 - 3u^2, -6uv, 6u), \quad \boldsymbol{p}_v = (6uv, 3v^2 - 3 - 3u^2, -6v), \\
&\boldsymbol{p}_{uu} = (-6u, -6v, 6), \quad \boldsymbol{p}_{uv} = \boldsymbol{p}_{vu} = (6v, -6u, 0), \\
&\boldsymbol{p}_{vv} = (6u, 6v, -6), \\
&\boldsymbol{e} = \frac{1}{\Delta}(2u, 2v, u^2 + v^2 - 1), \quad \text{where } \Delta = u^2 + v^2 + 1, \\
&E = 9\Delta^2, \quad F = 0, \quad G = 9\Delta^2, \quad EG - F^2 = 81\Delta^4, \\
&L = -6, \quad M = 0, \quad N = 6, \\
&K = \frac{-4}{9(u^2 + v^2 + 1)^4}, \quad H = 0.
\end{aligned}$$

Problem 2.3.4 We set $x = u,\ y = v,\ z = \log \frac{\cos v}{\cos u}$. Then

$$\boldsymbol{p}_u = (1, 0, \tan u), \quad \boldsymbol{p}_v = (0, 1, -\tan v),$$
$$\boldsymbol{p}_{uu} = (0, 0, \sec^2 u), \quad \boldsymbol{p}_{uv} = \boldsymbol{p}_{vu} = \boldsymbol{0}, \quad \boldsymbol{p}_{vv} = (0, 0, -\sec^2 v),$$
$$\boldsymbol{e} = \frac{1}{\Delta}(-\tan u, \tan v, 1), \quad \text{where } \Delta = \sqrt{\tan^2 u + \tan^2 v + 1},$$
$$E = \sec^2 u, \quad F = -\tan u \tan v, \quad G = \sec^2 v, \quad EG - F^2 = \Delta^2,$$
$$L = \frac{1}{\Delta}\sec^2 u, \quad M = 0, \quad N = -\frac{1}{\Delta}\sec^2 v,$$
$$K = -\frac{\sec^2 u \sec^2 v}{(\tan^2 u + \tan^2 v + 1)^2} = -\frac{\sec^2 x \sec^2 y}{(\tan^2 x + \tan^2 y + 1)^2}, \quad H = 0.$$

Problem 2.3.5 We only need to find points where the function

$$\lambda = \frac{L\xi^2 + 2M\xi\eta + N\eta^2}{E\xi^2 + 2F\xi\eta + G\eta^2}$$

is a constant $\sqrt{K}$, independent from (ξ, η). Thus we will determine points where

$$L = \sqrt{K}E, \quad M = \sqrt{K}F, \quad N = \sqrt{K}G.$$

By the calculation in Example 2.3.2 we get $M = 0$ and so $F = 0$. Hence if $|u| < \frac{\pi}{2}$, one of $\sin u = 0$, $\sin v = 0$ and $\cos v = 0$ holds. On the other hand, $L = \sqrt{K}E$ implies

$$\frac{abc}{\Delta} = \frac{abc}{\Delta^2}E.$$

Therefore we get

$$(*) \qquad \Delta = E.$$

Similarly $N = \sqrt{K}G$ implies

$$\frac{abc\cos^2 u}{\Delta} = \frac{abc}{\Delta^2}G.$$

Therefore we get

$$(**) \qquad \Delta\cos^2 u = G.$$

(1) In the case where $\sin u = 0$ we get $\cos^2 u = 1$ and

$$\Delta = \sqrt{b^2c^2\cos^2 v + c^2a^2\sin^2 v}, \quad E = c^2, \quad G = a^2\sin^2 v + b^2\cos^2 v.$$

On the other hand, $(*)$ and $(**)$ imply $E = G$. However it is a contradiction because $a > b > c$.

(2) In the case where $\sin v = 0$ we get $\cos^2 v = 1$ and

$$\Delta = \sqrt{b^2c^2\cos^2 u + c^2a^2\sin^2 u}, \quad E = a^2\sin^2 u + c^2\cos^2 u, \quad G = b^2\cos^2 u.$$

$(**)$ implies $b^2 = c^2\cos^2 u + a^2\sin^2 u$. Hence

$$\sin^2 u = \frac{b^2 - c^2}{a^2 - c^2}, \quad \text{i.e.,} \quad \sin u = \pm\sqrt{\frac{b^2 - c^2}{a^2 - c^2}}.$$

Then

$$\cos^2 u = \frac{a^2 - b^2}{a^2 - c^2}, \quad \text{i.e.,} \quad \cos u = \pm\sqrt{\frac{a^2 - b^2}{a^2 - c^2}}.$$

Substituting into $\boldsymbol{p}(u, v) = (a \cos u \cos v,\ b \cos u \sin v,\ c \sin u)$, we conclude that

$$\left(\pm a\sqrt{\frac{a^2 - b^2}{a^2 - c^2}},\ 0,\ \pm c\sqrt{\frac{b^2 - c^2}{a^2 - c^2}} \right)$$

are four umbilical points. (The above two ±'s are independent.)

(3) In the case where $\cos v = 0$ we get $\sin^2 v = 1$ and

$$\Delta = \sqrt{c^2 a^2 \cos^2 u + a^2 b^2 \sin^2 u}, \quad E = b^2 \sin^2 u + c^2 \cos^2 u, \quad G = a^2 \cos^2 u.$$

On the other hand, (∗) and (∗∗) imply $E = \frac{G}{\cos^2 u}$, that is, $b^2 \sin^2 u + c^2 \cos^2 u = a^2$. However it is a contradiction because $a > b > c$.

Since a coordinate (u, v) is applied when $|u| < \frac{\pi}{2}$ and $(0, 0, \pm c)$ is not contained in it (see Example 2.1.2), we need to verify that the points are not umbilical points separately. However at these points

$$K = \frac{c^2}{a^2 b^2}, \quad H = \frac{(a^2 + b^2)c}{2a^2 b^2},$$

that is,

$$\kappa_1 = \frac{c}{a^2}, \quad \kappa_2 = \frac{c}{b^2}$$

from Example 2.3.2.

Problem 2.3.6 This is a problem to determine the possible and impossible directions of $\boldsymbol{e}$. Since we calculated components of $\boldsymbol{e}$, we should use them. For the following (a) to (g), we can find it by looking at their graph in Sections 2.1 and 2.3, so only answers are given.

(a) The whole sphere is covered once.

(b) The whole sphere except for both north and south poles is covered once.

(c) The whole sphere except for the equator is covered once. The upper sheet of hyperboloid of two sheets covers the north hemisphere and the lower leaf of the sheet covers the south hemisphere.

(d) The north hemisphere is covered once. (The equator is not covered.)

(e) Similar to (d).

(f) The whole sphere is covered. The circle consisting of the highest points of a torus corresponds to the north pole, and the circle consisting of the lowest points of a torus corresponds to the south pole. The outside part surrounded by these two circles and the inside part surrounded by these two circles cover the whole sphere except for both poles just once respectively.

(g) The whole sphere except for both the north and the south poles is covered once.

(h) As we showed in the solution of Problem 2.3.2,

$$e = \frac{1}{\sqrt{u^2 + a^2}}(a \sin v, -a \cos v, u)$$

We will find For a given unit vector (ξ, η, ζ), in order to obtain u and v such that $e = (\xi, \eta, \zeta)$, we get

$$u = \pm \frac{a\zeta}{\sqrt{1 - \zeta^2}}$$

from $\frac{u}{\sqrt{u^2+a^2}} = \zeta$. Here we exclude the both poles because we need to assume $|\zeta| < 1$. When u is determined, v is determined uniquely with

$$\frac{a \sin v}{\sqrt{u^2 + a^2}} = \xi, \quad \frac{-a \cos v}{\sqrt{u^2 + a^2}} = \eta,$$

excluding integer multiples of 2π. When v changes to $v + 2\pi$, $x = u \cos v$ and $y = u \sin v$ do not change but $z = av + b$ does. That is, (u, v) and $(u, v + 2\pi)$ determine different points on the sphere. Hence the whole sphere except for both poles is covered infinite times.

(i) As we showed in the solution of Problem 2.3.3,

$$e = \frac{1}{u^2 + v^2 + 1}(2u, 2v, u^2 + v^2 - 1).$$

We need $\zeta < 1$ in order that $e = (\xi, \eta, \zeta)$ holds for a given unit vector (ξ, η, ζ). Thus we exclude the north pole. Then

$$\zeta = \frac{u^2 + v^2 - 1}{u^2 + v^2 + 1} = 1 - \frac{2}{u^2 + v^2 + 1}.$$

Thus

$$\frac{2}{u^2 + v^2 + 1} = 1 - \zeta.$$

By using this we obtain

$$\xi = \frac{2u}{u^2 + v^2 + 1} = u(1 - \zeta), \quad \eta = \frac{2v}{u^2 + v^2 + 1} = v(1 - \zeta).$$

Therefore

$$u = \frac{\xi}{1 - \zeta}, v = \frac{\eta}{1 - \zeta}.$$

Now we have found that the whole sphere except for the north pole is covered just once.

(j) As we showed in the solution of Problem 2.3.4,

$$e = \frac{1}{\Delta}(-\tan u, \tan v, 1),$$

where $\Delta = \sqrt{\tan^2 u + \tan^2 v + 1}$. We need $\zeta > 0$ in order that $e = (\xi, \eta, \zeta)$ holds for a given unit vector (ξ, η, ζ). Thus we exclude the south hemisphere and the equator. Then u, v are determined by

$$-\tan u = \frac{\xi}{\zeta}, \quad \tan v = \frac{\eta}{\zeta}$$

excluding integer multiples of π. As we know by definition equation, Scherk's surface is not defined at lattice points $x = \frac{\pi}{2} + m\pi,\ y = \frac{\pi}{2} + n\pi\ (m, n = 0, \pm 1, \pm 2, \ldots)$. In one lattice, for example $|x| < \frac{\pi}{2}$, $|y| < \frac{\pi}{2}$, the north hemisphere is covered once. See Chapter 5 for more details.

Problem 2.4.1 We replace indexes u, v of $\boldsymbol{p}_u$, $\boldsymbol{p}_v$ by 1, 2 and write (2.4.4) as

(i)
$$\boldsymbol{p}_i = \sum_{j=1}^{2} \boldsymbol{e}_j\, a_i{}^j.$$

By (2.4.12) we obtain

(ii)
$$\begin{aligned} d\boldsymbol{p}_i &= \sum_{j=1}^{2} d\boldsymbol{e}_j\, a_i{}^j + \sum_{j=1}^{2} \boldsymbol{e}_j\, da_i{}^j \\ &= \sum_{j,k=1}^{2} \boldsymbol{e}_k\, \omega_j{}^k\, a_i{}^j + \sum_{j=1}^{2} \boldsymbol{e}_j\, da_i{}^j + \sum_{j=1}^{2} \boldsymbol{e}_3\, \omega_j{}^3\, a_i{}^j. \end{aligned}$$

On the other hand, by (2.2.25) (and using (i) also) we get

(iii)
$$\begin{aligned} d\boldsymbol{p}_i &= \sum_{j,k=1}^{2} \boldsymbol{p}_k\, \Gamma_{ij}^k\, du^j + (*)\, \boldsymbol{e}_3 \\ &= \sum_{j,k,l=1}^{2} \boldsymbol{e}_l\, a_k{}^l\, \Gamma_{ij}^k\, du^j + (*)\, \boldsymbol{e}_3, \end{aligned}$$

where $(*)$ denotes one of L, M and N. Comparing coefficients of $\boldsymbol{e}_i$ $(i = 1, 2)$ in (ii) and (iii),

(iv)
$$\sum_{j,k=1}^{2} a_k{}^l\, \Gamma_{ij}^k\, du^j = \sum_{j=1}^{2} \omega_j{}^l\, a_i{}^j + da_i{}^j.$$

If we write it using the matrix notation, we obtain the equation to prove

(v)
$$A\Gamma = \omega A + dA.$$

Of course, we can write the above calculations (i) – (iv) using matrix notations.

Problem 2.5.1 $du = dr\, \cos\theta - r\, \sin\theta\, d\theta$, $dv = dr\, \sin\theta + r\, \cos\theta\, d\theta$. Using the rule (2.5.1) we have

$$\begin{aligned} du \wedge dv &= (dr\, \cos\theta - r\, \sin\theta\, d\theta) \wedge (dr\, \sin\theta + r\, \cos\theta\, d\theta) \\ &= r\, \cos^2\theta\, dr \wedge d\theta - r\, \sin^2\theta \wedge dr \\ &= r\, \cos^2\theta\, dr \wedge\, d\theta + r\, \sin^2\theta\, dr \wedge\, d\theta = r\, dr \wedge\, d\theta. \end{aligned}$$

Problem 2.6.1 Calculating $d\boldsymbol{p}$ by $\boldsymbol{p} = (a\cos u\, \cos v, a\cos u\, \sin v, a\sin u)$ and dividing it into the du-part and the dv-part, we obtain

$$\begin{aligned} d\boldsymbol{p} &= (-\sin u \cos v, -\sin u \sin v, \cos u)a\,du + (-\sin v, \cos v, 0)a \cos u\,dv \\ &= \boldsymbol{e}_1\, a\,du + \boldsymbol{e}_2\, a \cos u\,dv. \end{aligned}$$

Hence

$$\theta_1 = a\,du, \quad \theta_2 = a \cos u\,dv.$$

On the other hand, calculating $\boldsymbol{e}_3 = \boldsymbol{e}_1 \times \boldsymbol{e}_2$, we get

$$\boldsymbol{e}_3 = (-\cos u \cos v, -\cos u \sin v, -\sin u) = -\frac{1}{a}\boldsymbol{p}.$$

(We know $\boldsymbol{e}_3 = \pm\frac{1}{a}\boldsymbol{p}$ without calculation by the geometric fact that $\boldsymbol{e}_3$ is a unit normal vector and $\boldsymbol{p}$ is a vector with length a, orthogonal to the sphere.)

Next we calculate $d\boldsymbol{e}_1$, $d\boldsymbol{e}_2$, $d\boldsymbol{e}_3$ directly and we obtain

$$\begin{aligned} d\boldsymbol{e}_1 &= & &-\boldsymbol{e}_2 \sin u\,dv & &+\boldsymbol{e}_3\,du, \\ d\boldsymbol{e}_2 &= \boldsymbol{e}_1 \sin u\,dv & & & &+\boldsymbol{e}_3 \cos u\,dv, \\ d\boldsymbol{e}_3 &= -\boldsymbol{e}_1\,du & &-\boldsymbol{e}_2 \cos u\,dv. \end{aligned}$$

Hence

$$\begin{bmatrix} {\omega_1}^1 & {\omega_1}^2 & {\omega_1}^3 \\ {\omega_2}^1 & {\omega_2}^2 & {\omega_2}^3 \\ {\omega_3}^1 & {\omega_3}^2 & {\omega_3}^3 \end{bmatrix} = \begin{bmatrix} 0 & -\sin u\,dv & du \\ \sin u\,dv & 0 & \cos u\,dv \\ -du & -\cos u\,dv & 0 \end{bmatrix}.$$

Now

$$d{\omega_1}^2 = d(\sin u\,dv) = \cos u\,du \wedge dv = \frac{1}{a^2}(a\,du) \wedge (a \cos u\,du) = \frac{1}{a^2}\,\theta^1 \wedge \theta^2$$

and so we obtain $K = \frac{1}{a^2}$.

Moreover,

$${\omega_1}^3 = du = \frac{1}{a}\,\theta^1, \quad {\omega_2}^3 = \cos u\,dv = \frac{1}{a}\,\theta^2$$

gives

$$\begin{bmatrix} b_{11} & b_{12} \\ b_{21} & b_{22} \end{bmatrix} = \begin{bmatrix} \frac{1}{a} & 0 \\ 0 & \frac{1}{a} \end{bmatrix}.$$

We should note that although we calculate $\theta^1, \theta^2, {\omega_j}^i$ explicitly for a special $\boldsymbol{e}_1$, $\boldsymbol{e}_2$ in the previous calculus, the second fundamental form and K can be calculated easily for any orthonormal system $\boldsymbol{e}_1$, $\boldsymbol{e}_2$. As stated before, since $\boldsymbol{e}_3 = \boldsymbol{e}_1 \times \boldsymbol{e}_2$ is a unit normal vector, we have $\boldsymbol{e}_3 = \pm\frac{1}{a}\boldsymbol{p}$. By changing $\boldsymbol{e}_1$ and $\boldsymbol{e}_2$ if we need, we may assume $\boldsymbol{e}_3 = -\frac{1}{a}\boldsymbol{p}$ (i.e., $\boldsymbol{e}_3$ is a vector pointing inside the sphere). Then

$$d\boldsymbol{e}_3 = -\frac{1}{a}d\boldsymbol{p} = -\frac{1}{a}\,\theta^1\,\boldsymbol{e}_1 - \frac{1}{a}\,\theta^2\,\boldsymbol{e}_2,$$

and so

$${\omega_1}^3 = -{\omega_3}^1 = \frac{1}{a}\,\theta^1, \quad {\omega_2}^3 = -{\omega_3}^2 = \frac{1}{a}\,\theta^2.$$

Therefore we obtain $b_{11} = b_{22} = \frac{1}{a}$, $b_{12} = b_{21} = 0$. By (2.6.14) $K = b_{11}b_{22} - b_{12}b_{21} = \frac{1}{a^2}$. The above proof does not depend on the choice of $\boldsymbol{e}_1$ and $\boldsymbol{e}_2$.

Problem 2.6.2 Since $\boldsymbol{e}_1$ is a unit vector which is obtained by differentiating the equation (2.1.22) by u and dividing it by the length, we get

$$\boldsymbol{e}_1 = (-\sin u \cos v,\ -\sin u \sin v,\ \cos u).$$

Changing the roles of u and v, we similarly get

$$\boldsymbol{e}_2 = (-\sin v,\ \cos v,\ 0).$$

By a calculation similar to Problem 2.6.1 we get:

$$\begin{aligned}
&\boldsymbol{e}_3 = (-\cos u \cos v,\ -\cos u \sin v,\ -\sin u),\\
&\theta^1 = r\,du, \quad \theta^2 = (R + r\cos u)dv,\\
&\begin{bmatrix} {\omega_1}^1 & {\omega_1}^2 & {\omega_1}^3 \\ {\omega_2}^1 & {\omega_2}^2 & {\omega_2}^3 \\ {\omega_3}^1 & {\omega_3}^2 & {\omega_3}^3 \end{bmatrix} = \begin{bmatrix} 0 & -\sin u\,dv & du \\ \sin u\,dv & 0 & \cos u\,dv \\ -du & -\cos u\,dv & 0 \end{bmatrix},\\
&d{\omega_2}^1 = \cos u\, du \wedge dv = \frac{\cos u}{r(R + r\cos u)}\theta^1 \wedge \theta^2, \quad K = \frac{\cos u}{r(R + r\cos u)},\\
&\begin{bmatrix} b_{11} & b_{12} \\ b_{21} & b_{22} \end{bmatrix} = \begin{bmatrix} \dfrac{1}{r} & 0 \\ 0 & \dfrac{\cos u}{R + r\cos u} \end{bmatrix}.
\end{aligned}$$

Problem 2.6.3 First we write the first four equations in (2.2.25) together as

$$\boldsymbol{p}_{ij} = \sum_{k=1}^{2} \Gamma_{ij}^{k}\, \boldsymbol{p}_k + L_{ij}\, \boldsymbol{e} \quad (i, j = 1, 2).$$

The equation $\boldsymbol{0} = dd\boldsymbol{p}_i = \sum_{j,k=1}^{2} \boldsymbol{p}_{ijk}\, du^k \wedge du^j$ means nothing but $\boldsymbol{p}_{ijk} - \boldsymbol{p}_{ikj} = \boldsymbol{0}$. Expressing $\boldsymbol{p}_{ijk} - \boldsymbol{p}_{ikj}$ as a linear combination of $\boldsymbol{p}_i$, $\boldsymbol{p}_J$, $\boldsymbol{e}$, we consider the coefficient of $\boldsymbol{e}$ only. Since $d\boldsymbol{e}$ is a linear combination of only $\boldsymbol{p}_i$ by (2.2.25), we need not consider terms obtained by partial differentiation by u^k. Calculating carefully keeping in mind what was mentioned above, we get

$$\frac{\partial L_{ij}}{\partial u^k} - \frac{\partial L_{ik}}{\partial u^j} - \sum_{l=1}^{2} \Gamma_{ik}^{l}\, L_{lj} + \sum_{l=1}^{2} \Gamma_{ij}^{l}\, L_{lk} = 0.$$

Adding

$$\sum_{l=1}^{2} \Gamma_{kj}^{l}\, L_{il} - \sum_{l=1}^{2} \Gamma_{jk}^{l}\, L_{il}\ (= 0)$$

to this equation, we obtain what we want.

By studying tensor analysis, we may investigate the coefficient of $\boldsymbol{p}_l$ in $\boldsymbol{p}_{ijk} - \boldsymbol{p}_{ikj} = \boldsymbol{0}$, and obtain an equation which expresses K by Γ_{jk}^{i} and their partial differentials. From the above calculation, the readers can find advantages of learning the calculation methods using orthonormal frames.

Chapter 3

Problem 3.1.1 The first fundamental forms of catenoid and right helicoid are given by

$$d\boldsymbol{p} \cdot d\boldsymbol{p} = d\boldsymbol{q} \cdot d\boldsymbol{q} = (du)^2 + (u^2 + a^2)(dv)^2.$$

Although $\boldsymbol{q}$ gives a one-to-one correspondence from the (u, v)-plane to the right helicoid, $\boldsymbol{p}$ does not give a one-to-one correspondence from the (u, v)-plane to the catenoid because $\boldsymbol{p}(u, v) = \boldsymbol{p}(u, v + 2\pi)$. Hence it is not correct to say that the correspondence $\boldsymbol{q}(u, v) \longrightarrow \boldsymbol{p}(u, v)$ between two surfaces is isometric because $d\boldsymbol{p} \cdot d\boldsymbol{p} = d\boldsymbol{q} \cdot d\boldsymbol{q}$. (In this case we call the correspondence locally isometric.)

The second fundamental form of the catenoid is given by

$$-\frac{a}{u^2 + a^2}(du)^2 + a(dv)^2,$$

where we use the calculation in Example 3.3.6 and the second fundamental form of the right helicoid is given by

$$-\frac{2a}{\sqrt{u^2 + a^2}}\,dudv,$$

where we use the calculation in Problem 3.3.2. Both are minimal surfaces ($H = 0$) and they have

$$K = \frac{-a^2}{(u^2 + a^2)^2}, \quad \kappa_1 = \frac{a}{u^2 + a^2}, \quad \kappa_2 = \frac{-a}{u^2 + a^2}.$$

Problem 3.2.1 We write $u_x = \frac{\partial u}{\partial x}$, $u_y = \frac{\partial u}{\partial y}$, and so on. Then we have

$$du = u_x\,dx + u_y\,dy, \quad dv = v_x\,dx + v_y\,dy.$$

Thus

$$du^2 + dv^2 = ({u_x}^2 + {v_x}^2)dx^2 + 2(u_x u_y + v_x v_y)dxdy + ({u_y}^2 + {v_y}^2)dy^2.$$

Hence both (u, v) and (x, y) are isothermal coordinates if and only if

$$\text{(i)} \quad u_x u_y + v_x v_y = 0, \qquad \text{(ii)} \quad {u_x}^2 + {v_x}^2 = {u_y}^2 + {v_y}^2.$$

We rewrite (ii) as

$$\text{(ii)'} \quad {u_x}^2 - {u_y}^2 = {v_y}^2 - {v_x}^2$$

and add $2i\,(= 2\sqrt{-1})$ times (i) to (ii)', then we get

$${u_x}^2 + 2iu_x u_y - {u_y}^2 = {v_y}^2 - 2iv_x v_y - {v_x}^2.$$

Rewriting it as $(u_x + iu_y)^2 = (u_y - iv_x)^2$ and taking its square root, we get

$$u_x + iu_y = \pm(u_y - iv_x).$$

We should compare the real part and the imaginary part of both sides respectively. We note that the above equations are the Cauchy-Riemann equations when

$$w = u \pm iv, \quad z = x + iy.$$

Problem 3.2.2 Rewriting the given metric as

$$f(u)^2 \left(\frac{f'(u)^2 + g'(u)^2}{f(u)^2} dudu + dvdv \right),$$

and putting

$$\xi = \int \frac{\sqrt{f'(u)^2 + g'(u)^2}}{f(u)} du, \quad \eta = v,$$

we get

$$f(u)^2(d\xi d\xi + d\eta d\eta).$$

If we express u as a function of ξ, $f(u)^2$ becomes a function of ξ.

In the case of a catenoid and a right helicoid, ξ and η are given by

$$\xi = \int \frac{du}{\sqrt{u^2 + a^2}} = \sinh^{-1} \frac{u}{a}, \quad \eta = v.$$

Thus

$$u = a \sinh \xi, \quad u^2 + a^2 = a^2 \cosh^2 \xi.$$

and the given metric becomes

$$a^2 \cosh^2 \xi (d\xi d\xi + d\eta d\eta).$$

It is known that there exists an isothermal coordinate at least locally for a given Riemannian metric on a domain D in the (u, v)-plane not only in the case of surfaces of revolution but also of general surfaces. This is a result of partial differential equations theory, which is beyond this book's level.

Problem 3.2.3 Putting $\theta^1 = \sqrt{U + V}\, du$, $\theta^2 = \sqrt{U + V} dv$, we get

$$\begin{aligned} d\theta^1 &= \frac{V'}{2\sqrt{U + V}} dv \wedge du = -\frac{V'\, du}{2(U + V)} \wedge \theta^2, \\ d\theta^2 &= \frac{U'}{2\sqrt{U + V}} du \wedge dv = -\frac{U'\, dv}{2(U + V)} \wedge \theta^1. \end{aligned}$$

Comparing them with

$$d\theta^1 = -\omega_2{}^1 \wedge \theta^2, \quad d\theta^2 = -\omega_1{}^2 \wedge \theta^1 = \omega_2^1 \wedge \theta^1,$$

we get

$$\omega_2{}^1 = \frac{V' du - U' dv}{2(U + V)}.$$

Therefore

$$d\omega_2{}^1 = \frac{U'U' + V'V' - (U'' + V'')(U + V)}{2(U + V)^2} du \wedge dv$$
$$= \frac{U'U' + V'V' - (U'' + V'')(U + V)}{2(U + V)^3} \theta^1 \wedge \theta^2,$$
$$K = \frac{U'U' + V'V' - (U'' + V'')(U + V)}{2(U + V)^3}.$$

Problem 3.2.4 Rewriting the given metric to

$$ds^2 = \frac{(du - 2v dv)^2}{4(u - v^2)} + (dv)^2$$

and setting

$$\theta^1 = \frac{du - 2v dv}{2\sqrt{u - v^2}}, \quad \theta^2 = dv,$$

we have $d\theta^1 = d\theta^2 = 0$. In fact

$$\theta^1 = d(\sqrt{u - v^2}).$$

Hence $\omega_2{}^1 = 0, K = 0$. This is also clear because if we set

$$\xi = \sqrt{u - v^2}, \quad \eta = v$$

then we obtain

$$ds^2 = d\xi d\xi + d\eta d\eta.$$

Problem 3.3.1 We have

$$X(Yf) = \sum_{i=1}^{2} \xi^i \frac{\partial}{\partial u^i} \left(\sum_{j=1}^{2} \eta^j \frac{\partial f}{\partial u^j} \right)$$
$$= \sum_{i,j=1}^{2} \xi^i \frac{\partial \eta^j}{\partial u^i} \frac{\partial f}{\partial u^j} + \sum_{i,j=1}^{2} \xi^i \eta^j \frac{\partial^2 f}{\partial u^i \partial u^j}$$

and similarly we have

$$Y(Xf) = \sum_{j,i=1}^{2} \eta^i \frac{\partial \xi^j}{\partial u^i} \frac{\partial f}{\partial u^j} + \sum_{i,j=1}^{2} \xi^j \eta^i \frac{\partial^2 f}{\partial u^i \partial u^j}.$$

Hence we obtain

$$X(Yf) - Y(Xf) = \sum_{j=1}^{2} \sum_{i=1}^{2} \left(\xi^i \frac{\partial \eta^j}{\partial u^i} - \eta^i \frac{\partial \xi^j}{\partial u^i} \right) \frac{\partial f}{\partial u^j},$$

$$[X, Y] = \sum_{j=1}^{2} \left\{ \sum_{i=1}^{2} \left(\xi^i \frac{\partial \eta^j}{\partial u^i} - \eta^i \frac{\partial \xi^j}{\partial u^i} \right) \right\} \frac{\partial}{\partial u^j}.$$

It is enough to write

$$[[X,Y],Z]f = [X,Y]\cdot Zf - Z\cdot[X,Y]f$$
$$= X\cdot Y\cdot Zf - Y\cdot X\cdot Zf - Z\cdot X\cdot Yf + Z\cdot Y\cdot Xf$$

to prove the Jacobi equation.
Problem 3.3.2 (a) Putting $\varphi = \sum_{i=1}^{2} a_i\, du^i$, $\psi = \sum_{i=1}^{2} b_i\, du^i$, we have

$$\varphi\wedge\psi = (a_1b_2 - a_2b_1)\, du^1\wedge du^2$$
$$(\varphi\wedge\psi)(X,Y) = (a_1b_2 - a_2b_1)(\xi^1\eta^2 - \xi^2\eta^1)$$
$$= \begin{vmatrix} a_1 & a_2 \\ b_1 & b_2 \end{vmatrix}\begin{vmatrix} \xi^1 & \eta^1 \\ \xi^2 & \eta^2 \end{vmatrix} = \begin{vmatrix} \sum a_i\xi^i & \sum a_i\eta^i \\ \sum b_i\xi^i & \sum b_i\eta^i \end{vmatrix}$$
$$= \varphi(X)\psi(Y) - \psi(X)\varphi(Y).$$

(b) Since

$$d\varphi = \left(\frac{\partial a_2}{\partial u^1} - \frac{\partial a_1}{\partial u^2}\right) du^1\wedge du^2,$$

we get

$$(d\varphi)(X,Y) = \left(\frac{\partial a_2}{\partial u^1} - \frac{\partial a_1}{\partial u^2}\right)(\xi^1\eta^2 - \xi^2\eta^1),$$
$$X(\varphi(Y)) = \sum_{i,j=1}^{2} \frac{\partial a_i}{\partial u^j}\,\xi^j\,\eta^i + \sum_{i,j=1}^{2} a_i\,\xi^j\,\frac{\partial\eta^i}{\partial u^j},$$
$$Y(\varphi(X)) = \sum_{i,j=1}^{2} \frac{\partial a_i}{\partial u^j}\,\xi^i\,\eta^j + \sum_{i,j=1}^{2} a_i\,\eta^j\,\frac{\partial\xi^i}{\partial u^j},$$

and

$$\varphi([X,Y]) = \sum_{i,j=1}^{2} a_i\left(\xi^j\frac{\partial\eta^i}{\partial u^j} - \eta^j\frac{\partial\xi^i}{\partial u^j}\right),$$

where we should refer the solution of Problem 3.3.1 for the last equality. Now we get easily the given formula.
Problem 3.4.1 (a) Since $\theta^1 = \frac{1}{v}du$, $\theta^2 = \frac{1}{v}dv$, we get $d\theta^1 = \frac{1}{v^2}du\wedge dv$, $d\theta^2 = 0$. Putting ${\omega_2}^1 = -{\omega_1}^2 = -\frac{1}{v}du$, we get $d\theta^1 = -{\omega_2}^1\wedge\theta^2$, $d\theta^2 = -{\omega_1}^2\wedge\theta^1$.

(b) Let $\xi^1\boldsymbol{e}_1 + \xi^2\boldsymbol{e}_2$ be a parallel vector field along a given curve. Then we have

$$0 = \frac{d\xi^1}{dt} + \xi^2\frac{{\omega_2}^1}{dt} = \frac{\xi^1}{dt} - \xi^2\frac{1}{v(t)}\frac{du}{dt} = \frac{d\xi^1}{dt} + \xi^2,$$
$$0 = \frac{d\xi^2}{dt} + \xi^1\frac{{\omega_1}^2}{dt} = \frac{d\xi^2}{dt} + \xi^1\frac{1}{v(t)}\frac{du}{dt} = \frac{d\xi^2}{dt} - \xi^1.$$

We delete $\frac{d\xi^2}{dt}$ by using the equation obtained by differentiating the first equation and the second equation. Then we get

$$\frac{d^2\xi^1}{dt^2} + \xi^1 = 0.$$

Thus

$$\xi^1 = \alpha \cos t + \beta \sin t, \quad \xi^2 = \alpha \sin t - \beta \cos t,$$

where α, β are constants. For cases of $\alpha = 1, \beta = 0$ and $\alpha = 0, \beta = -1$ see Figure A.8 and A.9.

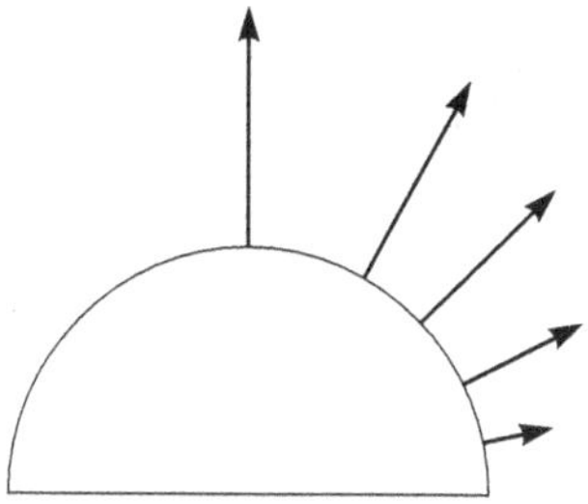

$\alpha=1,\ \beta=0$

Fig. A.8

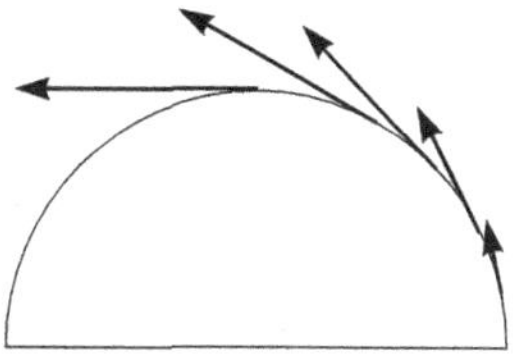

$\alpha=0,\ \beta=1$

Fig. A.9

(c) Let $\xi^1 \boldsymbol{e}_1 + \xi^2 \boldsymbol{e}_2$ be a parallel vector field along a given curve, then we have

$$\frac{d\xi^1}{dt} - \frac{1}{a}\xi^2 = 0, \quad \frac{d\xi^2}{dt} + \frac{1}{a}\xi^1 = 0$$

with calculations similar to (b). Deleting ξ^2 from these equations, we get

$$\frac{d^2\xi^1}{dt^2} + \frac{1}{a^2}\xi^1 = 0.$$

Therefore

$$\xi^1 = \alpha \cos \frac{t}{a} + \beta \sin \frac{t}{a}, \quad \xi^2 = -\alpha \sin \frac{t}{a} + \beta \cos \frac{t}{a},$$

where α, β are constants.

Problem 3.4.2 Setting

$$\frac{\partial}{\partial u^i} = \sum_{j=1}^{2} a_i^{\ j} \, \boldsymbol{e}_j$$

as in Section 2.4 of Chapter 2 (see (2.4.4)), we get

$$X = \sum_{i=1}^{2} \xi^i \frac{\partial}{\partial u^i} = \sum_{j=1}^{2} \eta^j \, \boldsymbol{e}_j,$$

where $\eta^j = \sum_{i=1}^{2} \xi^i a_i^{\ j}$. By (3.4.4),

$$\begin{aligned}\frac{DX}{dt} &= \sum_{j=1}^{2} \left(\frac{d\eta^j}{dt} + \sum_{k=1}^{2} \eta^k \frac{\omega_k^{\ j}}{dt} \right) \boldsymbol{e}_j \\ &= \sum_{j=1}^{2} \left\{ \sum_{i=1}^{2} \left(\frac{d\xi^i}{dt} a_i^{\ j} + \xi^i \frac{da_i^{\ j}}{dt} \right) + \sum_{i,k=1}^{2} \xi^i a_i^{\ k} \frac{\omega_k^{\ j}}{dt} \right\} \boldsymbol{e}_j.\end{aligned}$$

Now we use the solution of Problem 2.4.1 and get

$$\frac{DX}{dt} = \sum_{j=1}^{2}\left(\sum_{i=1}^{2}\frac{d\xi^i}{dt}a_i{}^j + \sum_{i,k,l=1}^{2}\xi^k\,\Gamma^i_{kl}\,\frac{du^l}{dt}\,a_i{}^j\right)\boldsymbol{e}_j$$
$$= \sum_{i=1}^{2}\left(\frac{d\xi^i}{dt} + \sum_{k,l=1}^{2}\xi^k\,\Gamma^i_{kl}\,\frac{du^l}{dt}\right)\frac{\partial}{\partial u^i}.$$

We can get the above equation directly if we go back to (2.2.25) and calculating in a similar way to (3.4.4).

Problem 3.5.1 Let X be a tangent vector field of a curve represented by arc length s. Then we have

$$X = \frac{du}{ds}\frac{\partial}{\partial u} + \frac{dv}{ds}\frac{\partial}{\partial v} = \frac{dt}{ds}\left(\frac{du}{dt}\frac{\partial}{\partial u} + \frac{dv}{dt}\frac{\partial}{\partial v}\right) = \frac{dt}{ds}\tilde{X}.$$

Hence

$$\mathbf{0} = \frac{DX}{ds} = \frac{D}{ds}\left(\frac{dt}{ds}\tilde{X}\right) = \frac{d^2t}{ds^2}\tilde{X} + \left(\frac{dt}{ds}\right)^2\frac{D\tilde{X}}{dt}.$$

Putting

$$\lambda = \left(\frac{dt}{ds}\right)^{-2}\frac{d^2t}{ds^2},$$

we get the equation that has to be proved. Conversely, if

$$\frac{D\tilde{X}}{dt} + \lambda\tilde{X} = \mathbf{0}$$

for a suitable λ, we get

$$\frac{DX}{ds} + \mu X = \mathbf{0}$$

with calculations similar to the above. Here μ is a function obtained by λ but we need not to know it explicitly. By $|X|^2 = \langle X, X\rangle = 1$ we get

$$0 = \frac{d}{ds}|X|^2 = 2\langle\frac{DX}{ds}, X\rangle,$$

that is, $\frac{DX}{ds}$ and X are orthogonal to each other. Hence we obtain

$$0 = \langle\frac{DX}{ds} + \mu X, X\rangle = \langle\frac{DX}{ds}, X\rangle + \mu\langle X, X\rangle = \mu.$$

Problem 3.5.2 From Problem 3.4.2 we have

$$X = \sum_{i=1}^{2}\frac{du^i}{ds}\frac{\partial}{\partial u^i},$$

thus, it is enough to set

$$\xi^i = \frac{du^i}{ds}.$$

Problem 3.5.3 If $\boldsymbol{p}(s)$ is a geodesic on the sphere, from $\boldsymbol{k}_g = \boldsymbol{0}$, we have $\boldsymbol{p}''(s) = \boldsymbol{k}_n$. Thus, $\boldsymbol{p}''(s)$ is orthogonal to the sphere. Now we consider $\boldsymbol{p}(s)$ as the space curve and write its Frenet-Serret's formula as:

$$\boldsymbol{e}_1 = \boldsymbol{p}'(s), \quad \boldsymbol{e}_1' = \boldsymbol{p}''(s) = \kappa\, \boldsymbol{e}_2, \quad \kappa > 0.$$

Since $\boldsymbol{p}''(s)$ is orthogonal to the sphere, it is a scalar multiple of $\boldsymbol{p}(s)$:

$$\kappa\, \boldsymbol{e}_2 = \boldsymbol{p}'' = \alpha \boldsymbol{p}.$$

Hence $\boldsymbol{e}_2 = \beta \boldsymbol{p}$, where $\beta = \frac{\alpha}{\kappa}$. On the other hand, $1 = |\boldsymbol{e}_2| = |\beta|\,|\boldsymbol{p}|$. Since $|\boldsymbol{p}|$ is a radius, β is a constant. Thus differentiating $\boldsymbol{e}_2 = \beta \boldsymbol{p}$, we get

$$\boldsymbol{e}_2' = \beta \boldsymbol{p}' = \beta \boldsymbol{e}_1.$$

Therefore the torsion $\tau = 0$ and it is a plane curve.

Problem 3.6.1 Putting $\theta^1 = du$, $\theta^2 = g dv$, we get

$$d\theta^1 = 0, \quad d\theta^2 = g_u\, du \wedge dv,$$

where $g_u = \frac{\partial g}{\partial u}$. Hence

$$\omega_2{}^1 = -g_u\, dv, \quad d\omega_2{}^1 = -\frac{g_{uu}}{g}\, \theta^1 \wedge \theta^2.$$

Therefore

$$K = -\frac{g_{uu}}{g}.$$

Problem 3.6.2 If we set

$$\boldsymbol{e}_1 = \frac{\partial}{\partial u}, \quad \boldsymbol{e}_2 = \frac{1}{g}\frac{\partial}{\partial v},$$

the tangent vector field of C is $\boldsymbol{e}_2$. C is a geodesic if and only if the differential equation (3.4.6) holds along C if we set $\xi^1 = 1$, $\xi^2 = 0$, $t = v$. By rewriting this by using $\omega_2{}^1$ which we calculated in Problem 3.6.1, we get $g_u(0, v) = 0$.

Next, for a constant K, integrating $g_{uu} = -Kg$ obtained in Problem 3.6.1, we obtain the following.

(i) When $K = 0$, $g(u, v) = \alpha(v) + \beta(v)u$.
(ii) When $K > 0$, $g(u, v) = \alpha(v) \cos \sqrt{K}u + \beta(v) \sin \sqrt{K}u$.
(iii) When $K < 0$, $g(u, v) = \alpha(v) \cosh \sqrt{K}u + \beta(v) \sinh \sqrt{K}u$.

By the initial condition $g(0, v) = 1$, $g_u(0, v) = 0$, we get $\alpha(v) = 1$, $\beta(v) = 0$. Therefore

(i) When $K = 0$, $ds^2 = (du)^2 + (dv)^2$.
(ii) When $K > 0$, $ds^2 = (du)^2 + \cos \sqrt{K}u\, (dv)^2$.
(iii) When $K < 0$, $ds^2 = (du)^2 + \cosh \sqrt{K}u\, (dv)^2$.

Chapter 4

Problem 4.1.1 (i)

$$
\begin{aligned}
d\varphi &= \frac{\partial}{\partial u}\left(\frac{u}{u^2+v^2}\right) du \wedge dv - \frac{\partial}{\partial v}\left(\frac{v}{u^2+v^2}\right) dv \wedge du \\
&= \frac{u^2+v^2-2u^2}{(u^2+v^2)^2}\, du \wedge dv - \frac{u^2+v^2-2v^2}{(u^2+v^2)^2}\, dv \wedge du \\
&= \left[\frac{u^2+v^2-2u^2}{(u^2+v^2)^2} + \frac{u^2+v^2-2v^2}{(u^2+v^2)^2}\right] du \wedge dv = 0.
\end{aligned}
$$

(ii) By using polar coordinates $u = r\cos\theta,\ v = r\sin\theta$,

$$
du = \cos\theta\, dr - r\sin\theta\, d\theta, \quad dv = \sin\theta\, dr + r\cos\theta\, d\theta,
$$

and substituting them into φ, we get $\varphi = d\theta$. Hence the integration φ along the unit circle in the positive direction is 2π. This does not contradict with Stokes' theorem. (Because φ is not defined at the origin.)

(iii) If we assume $df = \varphi$, the integration of φ along a circle must be 0, which contradicts (ii). Hence there is no such function f. This does not contradict Poincaré's lemma (Theorem 2.5.1).

Problem 4.2.1 If we assume that two geodesics intersect at P, Q like Figure 4.2.12 of Chapter 4, let A be the domain bounded by these geodesics and apply the formula (4.2.24). We obtain

$$
\int_A K\,\theta^1 \wedge \theta^2 = 2\pi - \varepsilon_{\mathrm{P}} - \varepsilon_{\mathrm{Q}},
$$

where $\varepsilon_{\mathrm{P}},\ \varepsilon_{\mathrm{Q}}$ are outer angles at P, Q. Since $K \le 0$, the left-hand side is negative or 0. On the other hand, outer angles are less than or equal to π (If the outer angle $\varepsilon_{\mathrm{P}} = \pi$, the two geodesics are tangent to each other at P. However, this is impossible since any geodesic is uniquely determined by a direction at 1 point). Thus $\varepsilon_{\mathrm{P}} < \pi,\ \varepsilon_{\mathrm{Q}} < \pi$ and so the right-hand side of the above equation is positive, which is a contradiction.

Problem 4.2.2 If we use the notation of Example 3.4.1, the tangent part of $\frac{de_1}{dv}\frac{dv}{ds}$ is the geodesic curvature vector $\boldsymbol{k}_g$. By (4.4.16) of Example 3.4.1 we get

$$
\kappa_g = \sin u\, \frac{dv}{ds}.
$$

Integrating κ_g along $C(u)$ which is the latitude line obtained by keeping u fixed, we get

$$
\int_{C(u)} \kappa_g\, ds = \int_0^{2\pi} \sin u\, dv = 2\pi \sin u.
$$

On the other hand, in order to calculate the domain A which is the northern part of the latitude line $C(u)$ we first rewrite the fundamental form as

$$
a^2(du)^2 + a^2\cos^2 u\,(dv)^2 = \theta^1\theta^1 + \theta^2\theta^2,
$$

where $\theta^1 = a\,du$, $\theta^2 = a\cos u\,dv$, and get the integration

$$
\int_A \theta^1 \wedge \theta^2 = \iint_A a^2 \cos u\, du \wedge dv = a^2 \int_0^{2\pi} dv \int_u^{\frac{\pi}{2}} \cos u\, du = 2\pi a^2(1 - \sin u).
$$

Since $K = \frac{1}{a^2}$, we get

$$\int_A K\,\theta^1 \wedge \theta^2 = 2\pi(1 - \sin u).$$

Therefore

$$\int_A K\,\theta^1 \wedge \theta^2 + \int_{C(u)} \kappa_g\, ds = 2\pi(1 - \sin u) + 2\pi\, \sin u = 2\pi,$$

and we obtain (4.2.24).

Problem 4.2.3 We can prove this with arguments similar to which we induced (4.2.25) from (4.2.24). Otherwise, we join P_1 and P_3 by a curve, divide the quadrilateral into two triangles, apply (4.2.25) to each triangle and add.

Problem 4.3.1 We assume that S is divided into f quadrilaterals. Each quadrilateral has four vertexes and there are four quadrilaterals at each vertex, so the number of vertexes are also f. By applying the equation of Problem 4.2.3 to each quadrilateral and adding the equations, we get $\int_A K\,\theta^1 \wedge \theta^2 = 0$ since the sum of the interior angles is $f \cdot 2\pi$.

Otherwise, by directly counting the number of vertexes v and sides e, we get $v = f$ and $e = 2f$ (each quadrilateral has four sides and each side belongs to two quadrilaterals). Thus $\chi(S) = v - e + f = 0$.

Problem 4.3.2 Triangulate S, integrate $d\varphi$ on each triangle, and apply Stokes' theorem (4.1.11). Then the sum of the integrations of φ along the perimeter of each triangle is 0. (Can be proved in a similar way to that of the Gauss-Bonnet theorem in this Chapter.)

Problem 4.3.3 When X is divided by its length $|X| = \langle X, X\rangle^{\frac{1}{2}}$, it becomes a unit vector field, which we denote by $\boldsymbol{e}_1$. Since this surface is oriented, we can determine uniquely (with the positive direction) the unit tangent vector $\boldsymbol{e}_2$ orthogonal to $\boldsymbol{e}_1$. In this way we obtain a moving frame $\boldsymbol{e}_1$, $\boldsymbol{e}_2$ defined on S and we consider the structure equations

$$d\theta^i = -\sum_{j=1}^{2} \omega_j{}^i \wedge \theta^j \quad (i = 1, 2), \quad d\omega_2{}^1 = K\,\theta^1 \wedge \theta^2.$$

Since $\boldsymbol{e}_1$, $\boldsymbol{e}_2$ are defined on the whole surface S, $\theta^i, \omega_j{}^i$ are also defined on the whole surface S. By Problem 4.3.2 the integration of $d\omega_2{}^1$ on S vanishes. Therefore

$$\int_S K\,\theta^1 \wedge \theta^2 = 0.$$

By the Gauss-Bonnet theorem, the Euler characteristic of S is zero.

Postscript

Differential geometry is, as the name suggests, a study in which we apply the methods of calculus to solve problems of geometry. We can say that its history is as old as that of calculus. In fact, C. Huygens (1626–1695) and I. Newton (1642–1727) studied curvatures of plane curves, and G. Monge (1746–1818) started studies of differential geometry of space curves in the 1770s. The theory of surfaces progressed gradually through the efforts of J. Bernoulli (1667–1784), L. Euler (1707–1783), J. B. M. C. Meusnier (1754–1793), C. Dupin (1784–1873), a student of G. Monge, and others such as J. L. Lagrange (1736–1813), A. L. Cauchy (1789–1857). We can safely state, however, that modern differential geometry began with "Disquisitiones generales circa superficies curvas (General investigations of surfaces)," a book published by C. F. Gauss (1777–1855) in 1827. In this book, Gauss defined the curvature of a surface as the limit of the ratio of an area of small domain on its corresponding sphere to that on the given surface. He expressed the curvature by using the first and the second fundamental forms and after a lengthy calculation he obtained the following equation:

$$\begin{aligned}4(EG-F^2)^2K = {}& E\left[\frac{\partial E}{\partial v}\frac{\partial G}{\partial v} - 2\frac{\partial F}{\partial u}\frac{\partial G}{\partial v} + \left(\frac{\partial G}{\partial u}\right)^2\right]\\ &+F\left[\frac{\partial E}{\partial u}\frac{\partial G}{\partial v} - \frac{\partial E}{\partial v}\frac{\partial G}{\partial u} - 2\frac{\partial E}{\partial v}\frac{\partial F}{\partial v} - 2\frac{\partial F}{\partial u}\frac{\partial G}{\partial u} + 4\frac{\partial F}{\partial u}\frac{\partial F}{\partial v}\right]\\ &+G\left[\frac{\partial E}{\partial u}\frac{\partial G}{\partial u} - 2\frac{\partial E}{\partial u}\frac{\partial F}{\partial v} + \left(\frac{\partial E}{\partial v}\right)^2\right]\\ &-2(EG-F^2)\left[\frac{\partial^2 E}{\partial v^2} - 2\frac{\partial^2 F}{\partial u \partial v} + \frac{\partial^2 G}{\partial u^2}\right].\end{aligned}$$

He stated the famous theorem that says the curvature depends only on the first fundamental form. In the same article, Gauss proved the Gauss-Bonnet theorem for a special case, i.e., for a triangle whose sides are its geodesics. (In 1848, O. Bonnet (1819–1892) proved the theorem for a general case.) It is also Gauss who

S. Kobayashi, *Differential Geometry of Curves and Surfaces*,
Springer Undergraduate Mathematics Series,
https://doi.org/10.1007/978-981-15-1739-6

systematized the method to study surfaces by representing them with two parameters u and v. In the era of Gauss there were no accurate maps even in Europe. Gauss himself was involved in both theoretical study and actual practice of surveying. It is no a coincidence, of course, that his work on surveying was done around the same time as his work on differential geometry.

The next trigger for a major advance in differential geometry was the inaugural lecture of G. F. B. Riemann (1826–1866) "Über die Hypothesen, welche der Geometrie zugrunde liegen (On the hypotheses that lie at the foundations of geometry)" at Göttingen University in 1854. In that lecture, Riemann proposed to extend Gauss' theory of surfaces to n-dimensional manifolds. This was the beginning of what is called Riemannian geometry today.

Then in 1872, F. Klein gave an inaugural lecture "Vergleichende Betrachtung über neuere geometrische Forschungen (Comparative considerations of recent geometric researches)" at Erlangen University, where he set forth a point of view that geometry is the study of the properties which are invariant under a given transformation group. Accordingly, various geometries such as Euclidean geometry, projective geometry, affine geometry, etc. became to be understood in a undefined standpoint. Klein's lecture had further effects over many years under the name of Erlanger Programm (Erlangen program).

Subsequently, through the 1901 paper on tensor analysis by C. G. Ricci (1853–1925) and T. Levi-Civita (1873–1941) and the 1917 paper on parallel transformations on Riemannian manifolds by Levi-Civita, Riemannian geometry continued its gradual advance. Ever since A. Einstein (1879–1955) used tensor analysis based on the four-dimensional Riemannian metric of indefinite format to describe his general theory of relativity (1916), the study of Riemannian geometry in higher dimensions became increasingly popular.

In parallel with the above developments, H. Weyl (1885–1955) created affine differential geometry (1918) by extracting the notions of covariant derivatives and parallel translations from Riemannian geometry. He also introduced projective differential geometry by extracting the notion of geodesics and conformal differential geometry (1921) by focusing on the notion of angles. Weyl's works were succeeded by American mathematicians P. Eisenhart (1876–1965), O. Veblen (1880–1960), T. Y. Thomas (1899–1984) and others, and had a defining influence on differential geometry in the 1920s and 1930s.

E. Cartan (1869–1951) in France, on the other hand, developed affine, projective and conformal differential geometries in harmony with Klein's philosophy from 1923 through 1925.

In the study of surfaces, we discussed in this text, not only the method of using vector fields $\boldsymbol{p}_u$, $\boldsymbol{p}_v$ which depends on the coordinate system u, v, but also the method of using the orthonormal system $\boldsymbol{e}_1$, $\boldsymbol{e}_2$, $\boldsymbol{e}_3$ unrelated to u, v. The latter is the so-called moving frame method introduced by J. G. Darboux (1842–1917), who is known for his four volumes on surface theory "Théorie générale des surfaces." Cartan made extensive use of moving frames, created the method of differential form, and established important results in Lie group theory and differential geometry. We have learned in this text that a sphere has a positive constant curvature and that

Poincaré's upper half plane has a negative constant curvature. Cartan discovered a symmetric space which generalized the above results to higher dimensions. Symmetric spaces play an important role today in unitary representation theory and others. The differential form became an essential tool in the study of not only the topology of manifolds but also algebraic geometry, ever since G. de Rham (1903–1990) used it in his Ph.D. thesis (1931) to describe cohomology of manifolds. It can be said that the concept of moving frames is the original notion of fiber bundles, which is important in topology and other fields today.

In Chapter 4 of this book we proved the Gauss-Bonnet theorem for surfaces. Its generalization to higher dimensions was proved in 1925 by H. Hopf (1895–1971), to closed submanifolds of general dimension, independently by C. B. Allendoerfer (1911–1974) and W. Fenchel (1905–1988) in 1940. Based on their proof, Allendoerfer and A. Weil (1906–1988) proved the theorem in the case of general closed Riemannian manifolds in 1943. The simpler proof that S. S. Chern (1911–2004) presented in the following year includes the concept of transgression that later became important in topology. This field achieved remarkable progress through discovery of the Pontrjagin (1908–1988) class (1944) and that of the Chern class (1946), followed by the generalized Riemann-Roch theorem by F. Hirzebruch (1927–2012) and the index theorem by M. Atiyah (1929–2019) and I. M. Singer (1924–).

In Chapter 3, Section 3.6, we studied geodesics from the viewpoint of the variational method. This is an old research field together with the history of calculus of variations. In particular, the study of closed geodesics is still an active area of research because it is linked to the Morse (1892–1977) theory in dynamical systems ever since H. Poincaré (1854–1912) proved that there exists at least one closed geodesic on a convex surface.

In Chapter 5 we referred only to the basics, but the theory of minimal surfaces is also a research field that progressed together with the calculus of variations. It began with J. L. Lagrange, the father of calculus of variations, followed by G. Monge, J. B. M. C. Meusnier, O. Bonnet, B. Riemann, K. Weierstrass (1815–1897), H. A. Schwartz (1843–1921), E. Beltrami (1835–1900), J. A. Plateau (1801–1883) who demonstrated in 1973 a minimal surface by forming a soap film on a frame of a closed curve made by a wire, S. N. Bernstein (1880–1968), T. Radó (1985–1965), and J. Douglas (1897–1965) who solved the Plateau problem. The theory of minimal surfaces is still an active area of research.

Differential geometry has deep relations with topology as shown by the Gauss-Bonnet theorem and its generalizations. It is also connected to dynamical systems and ordinary differential equations through geodesics and others. Its relations with partial differential equations and complex function theory have deepened through research of minimal surfaces and others; it has also interacted with algebraic geometry through Kähler manifolds. Differential geometry has been developed by becoming closely connected with various fields of mathematics as shown above. We can say this is a natural consequence because most areas of mathematics are based on manifolds of some kind, and differential geometry is calculus on manifolds. Further, we have already mentioned that differential geometry is the most important tool in the theory of relativity. When we inquire about the origin of the success of differential geometry, we realize that it is found in the notion of curvatures introduced by Gauss and Riemann, which leads us to realize how great these two mathematicians were.

As seen from the character of differential geometry, it is necessary for us to learn differential geometry methods when we conduct research in various areas of mathematics. Likewise, it is important that differential geometers gain deep knowledge of other fields of mathematics and physics and conduct research with a broad perspective. The author wishes that readers who have completed this book will go on to study manifold theory and then mathematics more widely, including such areas as Riemannian geometry, function theory (of one and two variables), Lie group theory, and algebraic geometry.

Index

S. Kobayashi, *Differential Geometry of Curves and Surfaces*,
Springer Undergraduate Mathematics Series,
https://doi.org/10.1007/978-981-15-1739-6

Printed in the United States
by Baker & Taylor Publisher Services